AF702825

World University Library

The World University Library is an international series of books, each of which has been specially commissioned. The authors are leading scientists and scholars from all over the world who, in an age of increasing specialization, see the need for a broad, up-to-date presentation of their subject. The aim is to provide authoritative introductory books for university students which will be of interest also to the general reader. Publication of the series takes place in Britain, France, Germany, Holland, Italy, Spain, Sweden and the United States.

E. L. Schatzman

The Structure of the Universe

translated from the French by Patrick Moore

World University Library

McGraw-Hill Book Company
New York Toronto

© E. L. Schatzman 1968
Translation © George Weidenfeld and Nicolson Ltd 1968
Library of Congress Catalog Card Number 67-22981
Phototypeset by BAS Printers Limited, Wallop, Hampshire, England
Printed by Officine Grafiche Arnoldo Mondadori, Verona, Italy

Contents

1 Introduction

During the past few centuries, the limits of the known universe have been further and further extended. It was Copernicus who first revived the ancient idea that the Sun lies in the centre of the Solar System, and that the Earth is nothing more than an ordinary planet; since the stars do not seem to move appreciably compared with each other, it was reasonable to suppose that they must be extremely distant. Copernicus' great book *De Revolutionibus Orbium Cœlestium*, published in 1543, caused a complete change in the astronomical outlook. Before many more decades had passed, the old idea of an Earth-centred universe had been abandoned.

The first suggestions that there may be star-systems far beyond the boundaries of our own system, or Galaxy, were made in the 18th century. In 1750 Thomas Wright, an English instrument-maker who lived from 1711 to 1786, speculated about 'isolated systems' in space, though it must be admitted that his ideas were decidedly confused and unscientific. Five years later, the German philosopher Immanuel Kant put forward a more coherent theory, and his ideas were generally accepted for many years.

Various dim, hazy patches known as nebulæ are to be seen in the night sky; of these, the most famous is the Nebula in Andromeda, generally known as Messier 31 (M31) because it was the 31st object in a later catalogue of nebulæ and star-clusters drawn up by the French astronomer Charles Messier. Several nebulæ were listed by M. de Maupertuis as long ago as 1742. Kant regarded them as external systems, far beyond our Milky Way or Galaxy, and in this he was of course correct, even though proof could not then be obtained. Humboldt, the great explorer and scientist, called these systems 'island universes', a term which is still often used.

Throughout the 18th and 19th centuries the nebulæ were closely studied, and many more were found, notably by Sir William Herschel, possibly the greatest astronomical observer of all time, who was inclined to support Kant's 'island universe' theory even though he was reluctant to commit himself too firmly. However, the question had to remain open so long as it remained impossible

1 Circumpolar star-trails (exposure 2 hours 14 minutes). The photograph also shows the track of a meteor which happened to cross the field of view during the exposure. The short, bright trail near the centre of the picture is the Pole Star. The fact that a trail shows demonstrates that Polaris is not exactly at the polar point (though, admittedly, its distance from the celestial pole is less than 1 degree).

to measure the distances of the nebulæ. In 1838 F.W.Bessel, in Germany, measured the first star-distance, but his method could not be applied to much more remote objects, and less direct means of investigation were essential.

The situation changed during the years following 1885. The appearance of a nova, or temporary star, in the Andromeda Nebula was very significant; it reached naked-eye visibility for a brief period in 1885, and is now known to have been a supernova – that is to say, a tremendous stellar outburst during which a star 'explodes' and hurls much of its material away into space. Then in 1917 Ritchey, working at the Mount Wilson Observatory with the new 100-inch reflecting telescope, discovered a nova in the spiral nebula NGC 6946 (that is to say, No. 6946 in the New General Catalogue of nebulæ drawn up by J.L.E.Dreyer at Armagh Observatory). Examination of old photographic plates showed two further novæ in the Andromeda Galaxy in 1909. These discoveries indicated that the nebulæ were likely to be external systems. However, final proof was obtained by E.E. Hubble, following a study of the remarkable variable stars known as Cepheids.

Most stars (including our Sun) shine steadily, but there are some which fluctuate in brightness over short periods. It was found that with the variables, called Cepheids (because the brightest member of the group is the star Delta Cephei), the real luminosity is linked with the period of variation; the longer the period, the greater the luminosity. Therefore, the distance of a Cepheid may be calculated simply by observing its regular changes in brightness. In 1923 Hubble, again using the Mount Wilson 100-inch reflector, found Cepheids in the Andromeda Nebula, and it was at once clear that the Nebula could not possibly belong to our Galaxy. Its distance had to be measured in hundreds of thousands of light-years – a light-year being the distance travelled by a ray of light in one year – equivalent to almost 6 million million miles. We now know that the Andromeda Nebula is over 2,000,000 light-years away. The term 'nebula' for these external systems is rapidly becoming obsolete, since a true nebula is a gas-cloud in our own Galaxy, and the external objects are galaxies in their own right. Many of them, including Messier 31, are spiral in shape.

Another interesting development was that the galaxies were found to be receding from us. By spectroscopic methods, to be described below, the velocities of recession could be measured, and many measurements were made by Hubble and his colleague, M. Humason, following 1912. It was found that apart from Messier 31 and a few other galaxies now known to be members of our local group, the galaxies were moving away – and the speeds of recession increased for the more distant galaxies. This led on to the so-called Hubble-Humason Law, linking recession with distance. By 1938 it seemed that a reliable relationship had been found, and this was still adopted by Einstein in 1950, when he published the third edition of his work entitled *The Meaning of Relativity*. On the assumption that the whole universe is expanding, and that the rate of expansion has always been constant, it was found that the expansion must have begun some 1,800,000,000 years ago, which would presumably correspond to the age of the universe. (It is now clear that this figure is a gross underestimate.) The methods which Hubble used for this determination are given at the end of the book.

Between 1952 and 1957 it became obvious that the estimated distances of the galaxies, and hence their velocities of recession, were much too low. By 1957 the time-scale had been increased sevenfold from the value adopted in 1938. The original work was due to Walter Baade, working with the 200-inch Palomar reflector in California, who announced his results at the 1952 Congress of the International Astronomical Union. Baade found that there was an error in the Cepheid period-luminosity law – or, more accurately, that there are two different kinds of Cepheids, one type being considerably more luminous than the other. The error meant that instead of being a mere 900,000 light-years away, as had been thought, the distance of the Andromeda Galaxy must be at least 1,800,000 light-years, and there would have to be corresponding increases for all the other galaxies, so that the observable universe was more than twice as large as had been estimated. Since then, the accepted value for the distance of Messier 31 has been further increased, this time to 2,200,000 light-years.

Human nature is often strangely reluctant to recognise facts which will lead to major changes in outlook. For instance, the French astronomer Henri Mineur, in 1946, had noted the differences between the two types of Cepheid variables, but had imagined that there must be a serious error in his interpretation, and in his calculations he had simply taken an average value for the two types. In fact, Mineur had all the essential information in his hands, but had been unable to interpret it in the same way that Baade did seven years later.

Baade's results were soon firmly supported by other astronomers, and it became clear that Hubble's older methods would have to be re-examined with a critical eye to see which of them could be retained. In 1956 Humason, Mayall and Sandage, at Palomar, announced a revised increase in the distances of the galaxies, working this time from a re-calibration of the brilliancies of the galaxies themselves. Almost immediately after this, Sandage undertook a study of the individual stars in some of the galaxies, and yet another distance-increase resulted. However, it seems that the estimates are at last being made on a reliable basis, and it is probable that no major errors remain.

In the early 1930s it was discovered that radio waves can be received from space. These are collected by means of radio telescopes, which come in many designs: they are quite unlike optical telescopes, and do not produce visible pictures of the objects under study, but they can provide information which could not be collected in any other way, and since the end of the war, radio astronomy has become of vital importance. Improvements in techniques mean that it is now possible to find the positions of radio sources with reasonable accuracy. At an early stage it was found that the radio sources do not coincide with bright stars, but with objects of different kinds. Supernova remnants in our Galaxy emit radio waves, and there are various special galaxies which are

2 The Coelostat at the solar tower in Tokyo Observatory is typical of many others in the world. During recent years, intensive solar studies have been carried out by Japanese astronomers at Tokyo and elsewhere. (Photographed in 1966).

remarkably powerful in the radio range. Mills' catalogue (the 3-C or third Cambridge catalogue), published in 1962, lists about 1,200 sources of all kinds. Since radio telescopes can be made to a large size, very weak signals can be picked up, and indeed radio telescopes can penetrate further into the universe than optical telescopes.

Another remarkable discovery was made early in the 1960s. Some of the radio sources were identified with objects which looked optically very much like faint stars, but are now called quasi-stellar objects (QSOs) or quasars. If our present interpretations are correct, quasars are very remote and super-luminous – perhaps 100 times more luminous than ordinary galaxies. Much work remains to be done, but it does seem that the distances of some of the quasars exceed 6,000 million light-years.

When we turn to the theories of cosmology, or the past and future history of the universe, it is convenient to begin with a book entitled *Cosmologische Briefe*, published by the German astronomer Lambert in 1761. (It is interesting to note, in passing, that the French translation appeared in 1770, and was printed in Belgium by the clandestine press at Bouillon.) Lambert rejected the idea that the Earth lies at the centre of the universe; this, of course, was only to be expected. But in addition to this, he suggested that the Sun was an ordinary star in our Galaxy, and that the Galaxy itself was nothing more than a typical system or island universe, lying among other systems of island universes. Some of Lambert's ideas are now known to be wrong, but his work was of great importance, and it is certainly true that neither the Sun nor our own particular Galaxy is of any importance whatsoever in the universe considered as a whole.

Much later, Seeliger (1895) and Neumann (1896) developed cosmological theories, basing their ideas upon Newtonian mechanics. They had no choice; Einstein's theories, which modified the old Newtonian ideas, had not then appeared. But it has become very clear that Einsteinian relativity is of prime importance in cosmology, and it may be said that the new era opened in 1917, with the publication of his *Cosmological Considerations*.

The German scientist Mach had suggested that the inertia of a body must be determined entirely by the total distribution of mass in the universe. Einstein accepted this idea, and found that it could be satisfied by a model of the universe consisting of a spherical closed system, without, however, introducing any definite 'outer edge'. The equations of relativity theory were such that the mathematical requirements for such a universe could be satisfied. Then, in 1917, the Dutch astronomer W. de Sitter found that there was a further possibility, not considered by Einstein in his original book. De Sitter showed that one direct consequence of the equations of general relativity could be an empty but expanding universe. The idea of a universe devoid of matter seems, at first sight, to be quite opposed to the very ideas which had been put forward in the new theory of gravitation which Einstein had termed general relativity, and it was always clear that the De Sitter universe cannot correspond to reality, but theoretically it was of great significance. Five years later, Friedmann, one of the greatest of Russian mathematicians, showed that there are other non-static solutions of Einstein's equations – and it was Friedmann's results which gave the first clear indication of cosmological models containing matter in a state of expansion.

It would be very difficult to describe all the cosmological models which have been proposed during the past forty years or so, but something must be said about the work of Lemaître and of Eddington, which were of special importance. In 1927 Canon Lemaître, of Belgium, put forward his theory that the universe began at a set moment in the remote past, and that all the material was originally concentrated into what has been termed a 'primæval atom'; subsequently he summarised his theories in a book, *Théorie de Atome Primitif*. In 1930 Sir Arthur Eddington, of Britain, drew attention to Lemaître's theories, and made many original contributions of his own.

There can be no doubt that developments in astronomical theory have provided a powerful stimulus to cosmological research, and

3 The 250-ft paraboloid at the Jodrell Bank Radio Astronomy Observatory in Cheshire. This great instrument remains the largest fully-steerable radio telescope in operation, and it has been responsible for fundamental advances in research into radio astronomy. It has even been used occasionally for tracking satellites and space-probes, though this is certainly not the purpose for which it was designed, and it is not now used in this way. (Photographed in 1966).

the two branches of research are closely linked. In 1938, for instance, Bethe in America and von Weizsäcker in Germany independently found that the source of stellar energy is to be found in nuclear reactions taking place deep inside the stars; this meant that the Galaxy must be several thousands of millions of years old. The age of the Earth, too, was estimated more precisely, and by 1946 was taken to be at least 3,000 million years (the current estimate is 4,700 million years). Obviously, this did not fit in with the idea that the universe itself was only about 1,800 million years old, and in 1948 three astronomers at Cambridge University – H. Bondi, T. Gold and F. Hoyle – attempted to solve the discrepancy by means of a novel and most interesting theory, in which the universe was taken to be in a steady state, with fresh matter being continuously created out of nothingness. In their subsequent researches, the Cambridge group were led to introduce modifications into general relativity theory in order to fit into this steady-state hypothesis.

The first results of the work of the American astronomer Schwarzschild and his pupils in connection with the evolution of the stars were discussed at the Rome congress of the International Astronomical Union in 1952. This work led to a much more accurate determination of the ages of the majority of the stars, and further information was gathered steadily, leading to evidence in favour of objects perhaps 15,000, 20,000 or even 25,000 million years old. Here, too, there was an apparent contradiction of the same kind as that which had led to the steady-state theory of the universe; some of the stars appeared to be much too old to fit in with the estimated age of the universe as calculated from the present-day expansion rate. Difficulties of this kind serve to place even more emphasis upon the close relationship between astronomy and cosmology.

Another aspect of the cosmological problem concerns what is known as the singular state, that is, that time when all the material in the universe was concentrated into one very small area of space. Conditions of very great density must also involve remarkably high temperatures. When temperatures of several thousand million

4 The 72-in Tautenburg reflector. This has a complex optical system and can be used either as a conventional reflector, such as a Cassegrain, or as a Schmidt. It takes a great deal of time and work to change from one optical system to another in normal telescopes, but with the Tautenburg reflector the change from Cassegrain to Schmidt, or vice versa, takes only a few hours.

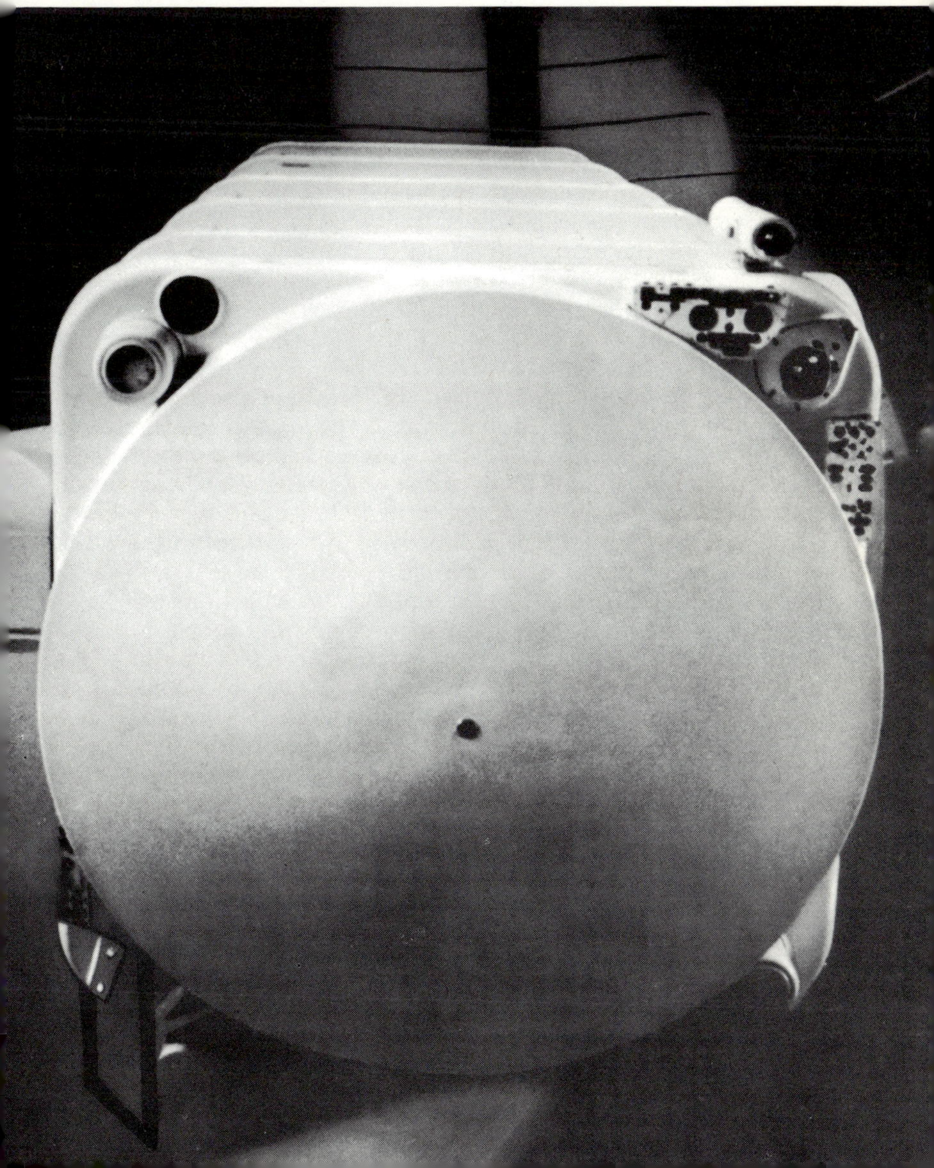

degrees are reached, an equilibrium is established, which determines the relative abundance of the nuclei of the various chemical elements, and the first attempts to explain the origin of the elements by equilibrium at very high temperature were made by Chandrasekhar and Henrich, but without real success. By the process of equilibrium alone, it seemed hopeless to satisfy the conditions under which the present-day relative abundances of the elements could have been reached. This difficulty led to the development of two different theories, each of which seemed to be outwardly plausible at the time. In 1950 Alpher, Bethe and Gamow suggested that the whole process of element-building, so to speak, had taken place during the first few seconds of the expansion of the universe; three years earlier, Bescow and Treffenberg, in Sweden, had proposed that the elements were originally formed in a sort of gigantic 'super-star' in gravitational equilibrium. Then, rather unexpectedly, the American astronomer Mayall discovered that some stars contain appreciable amounts of the element technetium – and this put a very different complexion upon matters, because this technetium could not be very old. It was already known that technetium is a radioactive element, with a half-life of 100,000 years; that is to say, in 100,000 years any given amount of technetium will have decayed by one-half. There was no escaping the conclusion that the technetium in stars was of comparatively recent formation, and the principle of uniformity, according to which all the elements were formed at the same time, had to be rejected.

The next step was to find out just how the various elements could be produced in a manner which would satisfy the relative abundances observed at the present time. Pioneer work in this field was undertaken by Margaret Burbidge, Geoffrey Burbidge, Fred Hoyle and W. Fowler, who in 1957 published a lengthy memoir in which the whole problem was critically examined. They came to the conclusion that element-formation was not due to one single event, but was caused by successive events operating all over the Galaxy, and involving what can only be called a 'primary substance' of unique composition. It must be said, however, that there is as yet no evidence in favour of such a fundamental substance. The researches of 'B2FH', as the four authors are often called in scientific circles, are not concerned with events which may have taken place during the first few seconds of the expansion of the universe. It is thought more likely that the answers to the problem are to be found inside the stars, or upon stellar surfaces.

Then, too, there have been other attempts to abandon the principles of uniformity which have influenced cosmological thought. For instance, Tolmann made a careful study of models of a non-uniform universe. Unfortunately, the difficulties involved in studies of models of this sort are so great that progress has been slight, even though the whole concept seems to be decidedly promising and will merit further investigation.

Broadly speaking, it is impossible to summarise the history of cosmology without bringing in the very recent discoveries – and progress is becoming more and more rapid, partly because of an improvement in instrumentation and technique, and partly because there are more and more astronomers and physicists available. There can be no doubt that cosmology has made more progress during the past twenty years than it did during the previous two hundred. Also, the link between astronomy, astrophysics and cosmology becomes closer and closer; we must deal with stellar distances, structure and evolution as well as with gravitational theories and relativity, and with atomic physics.

The present book is divided into four main parts. Part two gives a description of the Galaxy in which we live, together with the basic facts about the structure of the observable universe. The third part deals with 'time', both in measurement and in connection with the evolution of the universe. The fourth part describes the distribution of the galaxies in space, as well as their individual forms and characteristics. Lastly, consideration is given to the galaxies taken as a whole, in fact, to the universe itself.

5 The spiral galaxy M51 in Canes Venatici. The spiral arms are excellently marked and extend for a long way from the nucleus. Seen from a distance, our own Galaxy would be of this kind.

2 The Galaxy

On a dull day, with the sky overcast, it is difficult to draw up a proper picture of the universe which surrounds us; but on a dark night, such as often occurs in southern countries when the celestial objects are at their most brilliant, the universe looks fascinating indeed. But to allow one's imagination full rein is no obstacle to exact knowledge and detailed study, and, after all, Man's first ideas about the universe were drawn solely from observation. Theory had to come afterwards, and without observation it could not have come at all. Therefore, observational matters must be discussed first. To attempt to explain them, we must make full use of our knowledge of physics.

When considering the universe, the problems of matter, space, time and movement must be considered in their most fundamental aspects. The matter which surrounds us is seen in a variety of forms; it is distributed throughout space, it evolves in the course of time, and it is not motionless. The essential problem is to separate the different forms of matter, and to explain its movements, distribution and other characteristics. This cannot be tackled without first taking an overall view of what may be observed, and detailed analysis must follow this.

A short description of the universe

Naked-eye observation, telescopic observation, and (more important) photographic studies can tell us a great deal about the bodies to be seen in the universe. They can be classified, and it will be best to give an outline of this classification at the outset, so that the reader will become familiar with the various terms.

It is logical to begin with the stars, of which about 5,000 are visible to the naked eye over both northern and southern hemispheres of the sky. Telescopes increase this number considerably, and immense quantities of stars may be recorded on photographic plates. There is a great range in brilliancy, and the stars are divided into definite classes or *magnitudes* of apparent brightness. The fainter the star, the greater the magnitude. When two stars

6 The Milky Way.

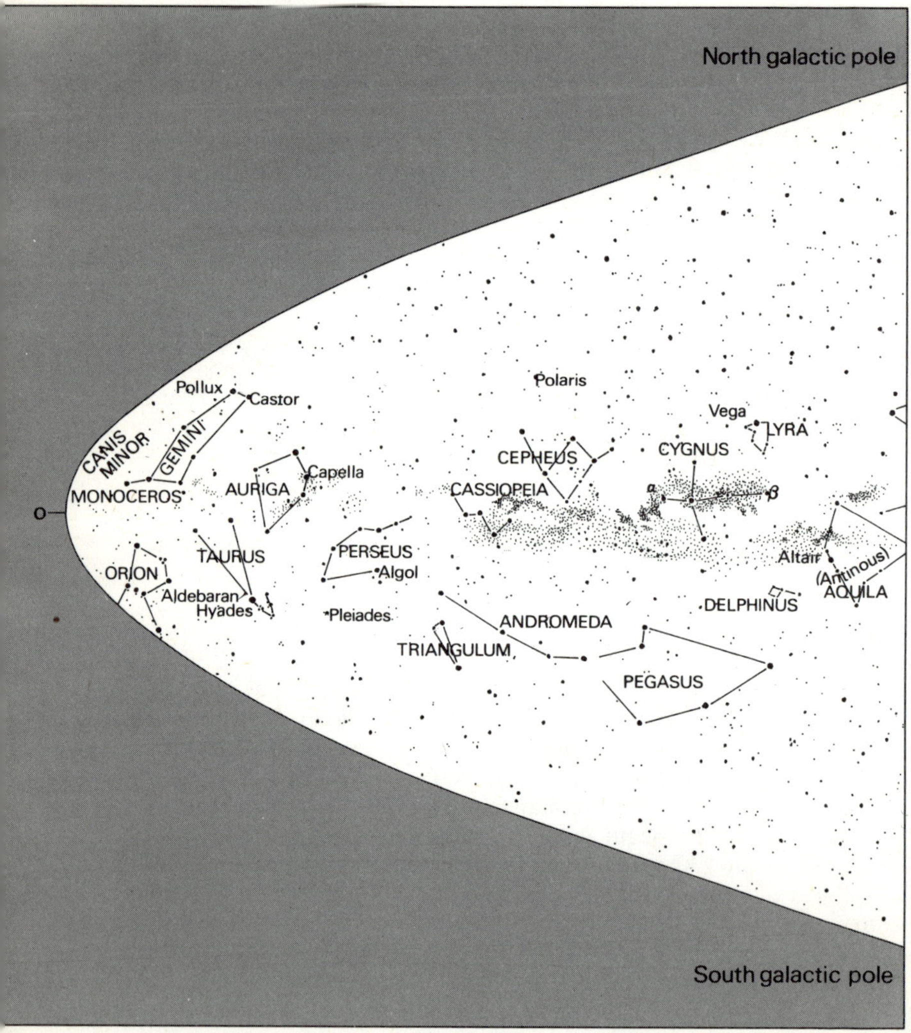

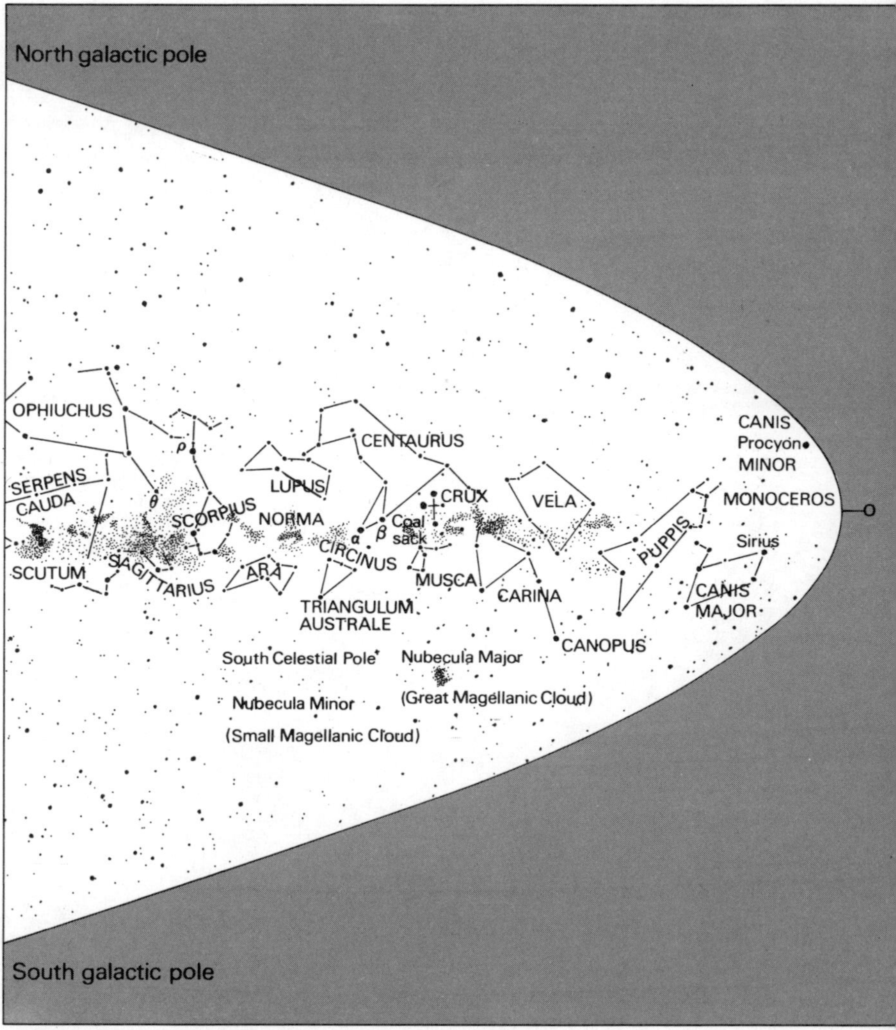
North galactic pole
OPHIUCHUS
SERPENS
CAUDA
SCUTUM
SAGITTARIUS
SCORPIUS
LUPUS
NORMA
ARA
CIRCINUS
CENTAURUS
CRUX
Coal sack
MUSCA
TRIANGULUM AUSTRALE
VELA
CARINA
CANOPUS
PUPPIS
CANIS MINOR
Procyon
MONOCEROS
Sirius
CANIS MAJOR
South Celestial Pole
Nubecula Major
(Great Magellanic Cloud)
Nubecula Minor
(Small Magellanic Cloud)
South galactic pole

7 Star-clouds in the Milky Way. This is a typical rich field showing vast numbers of stars together with nebulosity. It is the kind of view that can actually be seen by visual observation with an adequate telescope. The apparent crowding of the stars is due largely to line of sight effects.

8 The diagram shows the effects of parallax. Star E[1] is comparatively remote; star E is much closer to the Earth. E[1] seems to remain in the same position, but E shifts its position over an interval of six months because of the Earth's movement round the Sun. Measurement of the angle T_1ET_2 gives the star's parallax. In the case of Proxima Centauri, the angle is 1·6 seconds of arc, corresponding to a distance of slightly over four light-years.

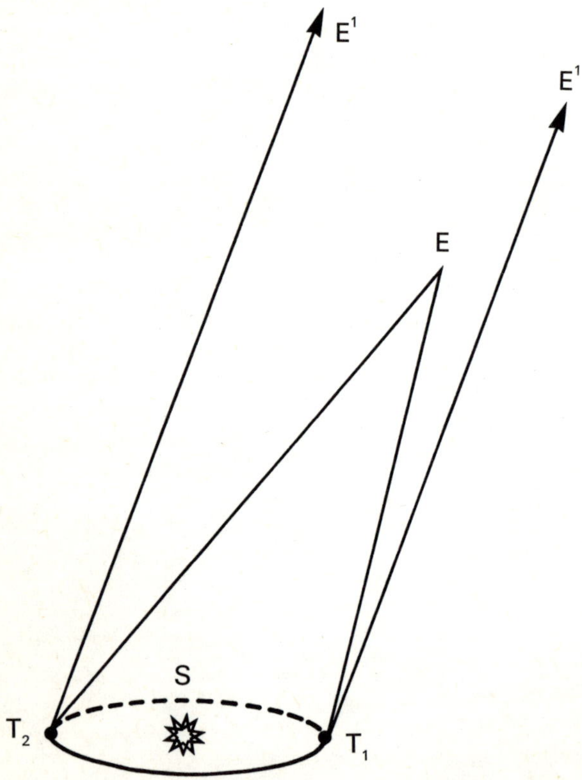

differ by five magnitudes, the fainter star has only 1/100 the brightness of the more brilliant – that is to say, a star of magnitude 2 is one hundred times as bright as a star of magnitude 7, and so on.

Sirius, the brightest star in the sky, is of magnitude −1·4, while the faintest star which can be photographed goes down to magnitude +23. There is therefore a range of almost 25 magnitudes, with a ratio of the order of 10,000,000,000 to 1. The photographic *Carte du Ciel*, which goes down to magnitude 13, contains about ten million stars, while the *Palomar Sky Atlas*, which extends to the 20th magnitude, includes about a thousand million stars.

It is obvious at a glance that the stars are unequally distributed in the sky. There is a great increase in star-density in the region of the Milky Way, the shining band which stretches across the sky and which has been known from the earliest times. Actually, the Milky Way is made up of a large number of faint stars which cannot be seen individually with the naked eye, so giving the impression of a glowing band. In other parts of the sky, well away from the Milky Way zone, there are far fewer stars.

A convenient method of studying star-density is to divide the sky into small equal areas, counting the number of stars visible in each; every unit is a square with its side measuring one degree on the celestial sphere. On the whole of the celestial sphere there are about 40,000 square degrees, and any preliminary chart constructed in this way shows up the unequal distribution. It also confirms the results of direct observation of the distribution of the stars in space, and the position of our Solar System in the Galaxy.

The Galaxy itself is made up of about 100,000 million stars arranged in the form of an immense disc, with a central condensation. When looking along the main plane of the disc, an observer sees a great many stars in roughly the same direction, and this produces the familiar Milky Way effect; in the perpendicular direction, the number of stars visible is much less.

Hazy, nebulous objects are found here and there among the stars. They are not stellar in aspect, but appear diffuse. The true nebulæ are clouds of gas mixed with dust; they may look bright,

9 The magnificent nebula M8 in Sagittarius. The gaseous nebula is seen in red hydrogen light. The luminosity is due to fluorescence by the excitation of atoms by ultra-violet radiation emitted by the star at the centre of the nebula. Other diffuse nebulosities are visible in the photograph.

because they are emitting light, or dark, because they absorb light and blot out stars lying beyond. The objects formerly known as extra-galactic nebulæ are in fact galaxies, in every way comparable with the Galaxy in which we live; they are enormous systems containing vast numbers of stars. The galaxies are of supreme importance, because they make up the essential building-blocks of the universe. The stars themselves are masses of gas, and they collect together into these huge systems which we call galaxies.

Distance determinations

When trying to draw up a distance-scale for the universe, the Solar System is the obvious starting-point. The Earth moves round the Sun in an orbit which has a radius of 93,000,000 miles, or 150,000,000 kilometres, so that in six months the Earth passes through positions which are separated by 300,000,000 kilometres. Originally, it was hoped that this displacement would produce an apparent shift in the positions of the nearer stars as against the more distant objects in the sky; and it is quite true that this shift, known as parallax, does occur. However, accurate instruments are needed for the parallax movements to be measured, and nothing of the sort could be done four centuries ago. Copernicus realised that the apparent lack of parallax shift was no obstacle to his theory that the Earth moves round the Sun – provided that the stars are extremely remote. Later in the sixteenth century, the great Danish astronomer Tycho Brahe put forward counter-arguments. He did not believe the stars to be immensely distant, and from this he went on to claim that the Earth must lie at rest in the centre of the universe. It was not until 1838 that the first successful parallax-measure was made, by F. W. Bessel in Germany.

The nearest known star, excluding the Sun, is Proxima Centauri in the southern hemisphere of the sky. Here, the apparent displacement due to parallax over a period of six months is 1·6 seconds of arc, so that an observer on Proxima would measure the diameter of the Earth's orbit round the Sun as less than 2 seconds

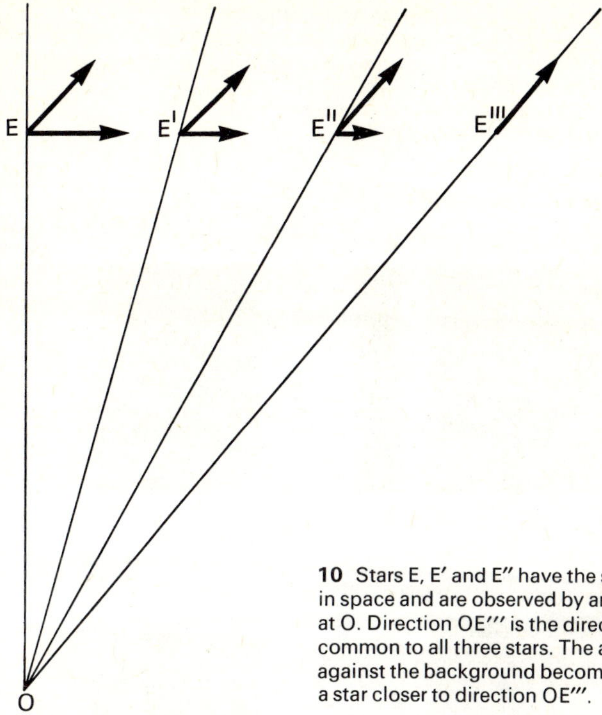

10 Stars E, E′ and E″ have the same velocity in space and are observed by an observer at O. Direction OE‴ is the direction of the velocity common to all three stars. The apparent velocity against the background becomes less for a star closer to direction OE‴.

of arc. This is very slight; an angle of 1 second is roughly the angle subtended by one centimetre seen from a distance of two kilometres. Proxima Centauri is about 250,000 times as far away as the Sun, and is not visible with the naked eye.

All measures of star-distances are based upon parallax displacements, but the actual method can be applied to only about 10,000 stars; with more distant objects (beyond about 20,000,000 times the distance of the Sun) the parallax shifts become so slight that they are swamped in the unavoidable errors of observation.

As has been noted, the light-year is a convenient unit for star-distance measures. On this reckoning, Proxima Centauri is rather over 4 light-years away, while the parallax method can extend out to objects at something like 300 light-years.

11 The Doppler effect, or Red Shift. With an approaching body the spectral lines are shifted toward the blue or short-wave end, whereas with a receding body the shift is to the red or long-wave end. The diagram is schematic, but shows the general principle, which is of fundamental importance in all stellar astronomy and cosmology.

Violet shift
approaching
object

Comparison
laboratory
spectrum

Red shift
receding
object

12 The radio telescope at the Tokyo Observatory. This is one of the smaller paraboloids, fully-steerable, and used mainly for solar work. (Photographed in 1966).

Absolute magnitude

The apparent magnitude of a star depends partly upon its distance and partly upon its real luminosity. The *absolute magnitude* may be defined as the apparent magnitude that a star would have if it were observed from a standard distance of 32·6 light-years, that is to say at a distance from which the radius of the Earth's orbit would subtend an angle of 0·1 seconds of arc. The absolute magnitude of the Sun is 4·62, so that it is neither particularly luminous nor particularly feeble.

Star-streaming

Another method which has been used successfully in measuring the distances of some more remote stars is based on the fact that there exist groups of stars which share a common movement in space relative to the Sun, and which show the phenomenon known as star-streaming. To show how this can be turned to good account, it is useful to draw an analogy with a train which is moving along a straight track. Objects which lie close to the train appear to shift more quickly than objects which are further away; as seen from the back of the train, the rails seem to meet at a point in the far distance, and while the train is in motion all objects appear to rush away along paths which converge toward the vanishing-point. On the other hand, an observer at the front of the train will have the impression of objects separating from each other and disappearing to either side of the advancing train. It is easy to see how this principle may be applied to the Sun, which may be said to stand for the moving train while the other stars represent objects such as trees and houses. Each group of stars seems to suffer displacement toward a single point, which is called the *convergent* of the stream; the direction in which this point lies is also the direction of the apparent movement of the star-group relative to the Sun. The distances of the stars in a stream are obtained from measuring their apparent shifts and their velocities along the line

13 Remains of the 72-in Rosse reflector at Birr Castle, Eire. This telescope was built by the third Earl of Rosse in 1845 and remained the largest reflector in the world for over half a century. With it Lord Rosse discovered the spiral structure of some of the galaxies, including M51. The mirror was of speculum metal – it was the last really large metal mirror to be built – and the mounting was so cumbersome that one could examine only a limited area of the sky to either side of the meridian. The telescope was last used in 1908, after which the mounting became unsafe and the whole structure was dismantled. The mirror is now displayed at the Science Museum in London; the tube remains at Birr Castle. (Photographed in June 1967).

of sight. For these latter measurements, use must be made of the famous Doppler effect.

The Doppler effect

The easiest way to explain the Doppler effect is to study the behaviour of sound waves. With a moving train which is sounding its whistle, the difference in the whistle-note between the approaching and the receding part of the line is very obvious; the note is high-pitched when the train approaches, and drops when the train moves away from the observer. In other words, the frequency is higher with approach, lower with recession, and the relative change in frequency is determined by the ratio of the speed of the source to the speed of propagation of the sound waves. A sound wave moves at about 330 metres per second, so that a train travelling at 120 kilometres per hour has a velocity 10 per cent of that of the speed of sound. The associated change in frequency corresponds to an increase of 10 per cent when the train is approaching the observer, and to a decrease of 10 per cent when the train is moving away – so that there is an overall difference of almost two tones.

Light, which is an electromagnetic vibration, shows the same sort of effect. With an approaching light source, the frequency is raised, while with a receding source the frequency is lowered. Since wavelength is inversely proportional to frequency, the receding source seems to have the longer wavelength, but of course the effect is much less obvious, because light waves move so much more rapidly than sound waves. At a speed for the observer of 30 kilometres per second (about 100,000 kilometres per hour), the wavelength is changed only by one ten-thousandth, which is not very much. All the same, a luminous source receding at half the velocity of light would show a Doppler effect to such a degree that its blue radiation would appear red to an observer lying at rest relative to the source. One amusing anecdote has been told in this connection. An American astronomer was brought before a court, accused of driving across a road despite the red traffic-light. He told the

14 A star-field photographed with the objective prism equipment at Fehrenbach. The field was photographed twice on the same plate, each star producing a small spectrum – blue on the right for the upper spectrum, blue on the left for the lower spectrum. This method makes it possible to identify the various spectral lines. The Hγ and Hδ lines of the upper spectrum are approximately opposite the Hγ and Hδ lines of the lower spectrum, but on closer examination it is seen that the apparent distance between the Hγ of the upper spectrum and Hδ of the lower varies from star to star. These differences are due to the fact that the stars have different radial velocities so that the Doppler shifts are not the same.

15 The dome of the 72-in Tautenburg telescope. This particularly modern type of dome houses the 'all-purpose' reflector shown in figure 4. (Photographed in 1963).

judge that he had seen the light as green, owing to the Doppler effect – but unfortunately for him, the judge knew just what was meant by the Doppler effect, so he waived the fine for jumping the traffic-light and fined the motorist for driving too fast! (Of course, to have seen the red light as green, the motorist would have had to have been driving at about one-third the velocity of light.)

When the speed of the source approaches the velocity of light, there is no straightforward way in which to calculate the change in apparent wavelength. To understand this, it is helpful to consider the relevant frequency, which is inversely proportional to the wavelength, and imagine that the observer is moving at the velocity of light. To him, the waves in a pencil of light-rays would appear completely stationary, because at each point he would always see the same wave – and consequently, the wavelength would be infinite! It follows that to an observer approaching the velocity of light, all wavelengths from a luminous source will increase indefinitely. The necessity of considering such effects, which are dealt with in the theory of special relativity, has recently been driven home by the discovery of the quasars. This subject will be discussed further in Sections 4 and 5 of the present book.

So far as the stars are concerned, Doppler effects are measured by means of spectroscopy. An astronomical spectroscope may be said to analyse the light coming from the stars, producing a coloured band or *spectrum* crossed by dark lines; each line is due to some particular element or group of elements. In the laboratory, these elements can be produced, and their wavelengths measured; the identifications are easy in many cases, particularly as the great physicist Kirchhoff showed, as long ago as 1859, that each element is capable of absorbing radiation of the same wavelength as that which it emits. If the spectral lines found with the stars are shifted slightly compared with the equivalent lines measured in the laboratory, the cause can lie only in a toward-or-away motion; in other words, the Doppler effect. The amount of Doppler effect is therefore a reliable way of estimating the real *radial movements* of the stars, and the same principles can be applied to the galaxies.

The distances of the stars in a stream

After this slight but necessary digression, let us return to the star-streams. For each star in a stream, it is possible to measure the *proper motion*, or apparent individual movement against the background of more distant stars, and also the *radial motion*, which is the toward-or-away movement as shown up by the Doppler effect. Take, for instance, a star of a stream which is seen at an angle of 45 degrees to its convergent. The longitudinal (radial) velocity and the transverse (proper motion) component are equal. If the velocity of the star in space is 42 km/sec, the radial velocity is 30 km/sec and the transverse velocity is also 30 km/sec, the apparent shift in position over a period of ten years will be 2·06 seconds of arc, so that its annual proper motion will be 0·206 seconds of arc. The distance of each star in the stream can then be worked out very accurately. This method has been applied successfully to an important group of stars in the constellation Taurus, known as the

Hyades cluster. The group lies at about 120 light-years from the Sun, and has a velocity, relative to the Sun, of 44 km/sec. By now, the distances of 159 individual stars in the Hyades have been determined with great precision, and this means that the absolute magnitudes can also be found. Therefore, the Hyades may be used as standards of reference for other groups, lying further away and consequently more difficult to measure directly.

Variable stars

As has been noted, there are some stars which show short-term variations in brilliancy, and of these the most important are the Cepheids, whose periods are short (from a few days up to several weeks) and whose behaviour is quite regular. In 1912, Miss Henrietta Leavitt, at Harvard, made a discovery which proved to be fundamental in the determination of stellar distances. She was studying Cepheids in the Small Magellanic Cloud, which is a southern-hemisphere object, and which is one of two minor galaxies associated with our own. Because the Cloud is relatively remote, all the various objects in it may be effectively regarded as being at the same distance from us, and Miss Leavitt found that the Cepheids with the longer periods were always the brightest – which meant that they must also be the more luminous. This was, of course, the first inkling of the Cepheid period-luminosity law. It is certainly valid, and it means that the distance of a Cepheid can always be worked out as soon as its period of variation is known. The only real problem is to find an absolute-magnitude scale. Unfortunately, all Cepheids are relatively distant, so that their parallax shifts are too small to be measured.

A related class, made up of variable stars with very short periods, has been found to be most useful in this connection. These stars are known as RR Lyræ variables, after the prototype example RR Lyræ, which fluctuates between magnitudes 7·1 and 7·8 in a period of 0·57 day. All RR Lyræ variables have approximately the same luminosity, so that all that needs be done is to measure their

16 The convergence of stars in Taurus. The positions of stars in the Hyades cluster are shown. Each star has a small arrow whose length is proportional to the star's annual proper motion. All the arrows converge toward the same point. Just as the lines of a railway track seem to meet near the horizon (assuming that the track is straight), so the convergence of the arrows shows that the stars have parallel paths in space. The direction toward which the arrows are converging is of course the direction of the stars.

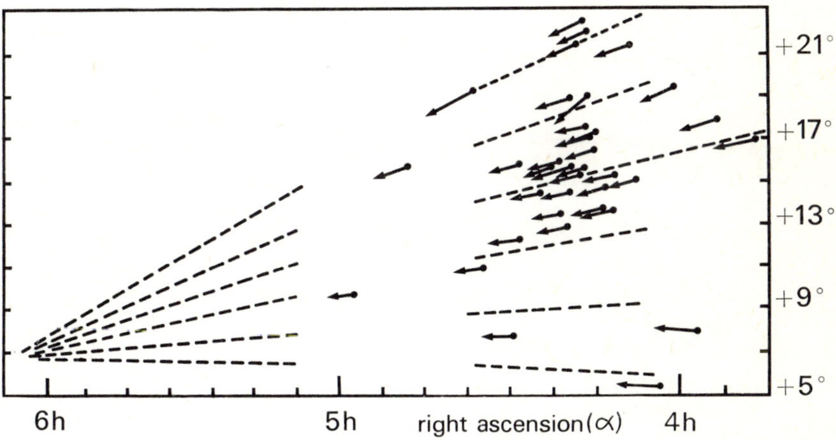

apparent magnitudes; since the luminosities are equal, the brighter stars will be the closer to us. As they are in motion relative to the Sun, a method much the same as that used in star-streaming can be applied to the RR Lyræ stars considered as a class. The direction of movement of the stars relative to the Sun is known. Stars lying in a direction perpendicular to the direction of their motion relative to the Sun will show proper motions against the celestial sphere, which can be measured; stars lying along the line of the Sun's relative motion, in either direction, will have measurable

17 The 98-in Isaac Newton reflector, taken in 1965, when the telescope was under construction at Newcastle-on-Tyne. The mirror is not in the tube, but the general pattern of the telescope stands out well. The telescope is now in operation at the Royal Observatory at Hurstmonceaux, in Sussex. It is by far the largest telescope in Britain, and the fifth largest in the world.

radial velocities. By comparing the mean proper motions and the mean radial velocities, and knowing that differences in apparent magnitudes must be due to differences in the real luminosities of the stars concerned, the absolute magnitudes of the RR Lyræ variables can be worked out very accurately. At the standard distance of 32·6 light-years, an RR Lyræ star would appear bright, with an apparent magnitude of 0, so that it is about one hundred times as luminous as the Sun.

Distance determinations for the Cepheids, which have longer periods and which are more powerful, can be carried out by much the same method, and here too the absolute magnitudes can be found; for instance, a Cepheid with a period of 100 days has an absolute magnitude of about −5. There is no real difficulty about studying Cepheids down to an apparent magnitude of +20 or so. It is known that a difference of 25 magnitudes corresponds to a distance ratio of 1:100,000. Therefore, if a star of absolute magnitude −5 appears to be of apparent magnitude +20, its distance must be 3,260,000 light-years.

This technique must depend upon a theory about the nature of the stars, because the statement that Cepheids of the same period have the same intrinsic brightness is tantamount to saying that they must have had a similar origin and have gone through the same phases in evolution. It is now known that variability is a characteristic which occurs for only a relatively short period during a star's career, a very long time after the star was formed; also, not all stars become variable at any particular stage. It is true to say that a star in the process of formation 'does not know' whether it will ever become a variable, or whether it will not; also, stars at an early stage in their evolution give us no clue as to whether they are likely to go through a variable stage. Yet it seems certain that Cepheids of equal period must be essentially similar to each other, and must have developed along the same lines.

18 The Large Magellanic Cloud. It is a galaxy visible from the southern hemisphere and appears to be a true companion of our own Galaxy.

The distribution of stars in the Galaxy

As has been noted, the stars in the Galaxy are arranged in the form of a disc; the Sun, of course, lies within this disc. In a direction along that of the plane of the disc, the number of stars per square degree is very great; at a direction right-angles to the main plane, the apparent number of stars per unit area is much less. To complete this description, something should be said about the distances of the stars; the short-period variables have been of immense importance in this connection, and it has been possible to show, with pleasing accuracy, that the Sun lies about 30,000 light-years from the centre of the Galaxy.

By means of absolute-magnitude determinations (notably by spectroscopic methods, of which more will be said later), and by straightforward counts, it has been possible to determine the distribution of the stars in space in the neighbourhood of the Sun. For all types of stars, the density-rate falls off with increasing distance from the plane of the Galaxy, as is only to be expected, but each variety of star has its own characteristic distribution with regard to the mean distance of the group from the galactic plane. Considering that most of the stars lie within a region only about 300 to 400 light-years in thickness, there is a wide range in the depths of zones occupied by the various star-types. For example, the blue dwarfs lie within about 150 light-years to either side of the plane. The Sun (a yellow dwarf) is situated almost exactly on the main plane.

The fact that the stars are not all distributed in the same way throughout the Galaxy may be explained quite simply in terms of elementary laws of celestial mechanics. What must be done first is to consider the forces which operate in the Galaxy as a whole.

Gravitation

In the latter part of the seventeenth century, Sir Isaac Newton showed that the attraction between two bodies is proportional to

19 The North American Nebula, NGC 7000 Cygni, photographed with the 48-in Schmidt at Palomar. This is contained in our Galaxy and is not an external system. Its nickname is derived from its shape, but long-exposure photographs are needed for it to be seen to advantage.

the product of their masses and inversely proportional to the square of the distance between them. Yet this force is surprisingly small. For instance, the Moon will attract an object on the Earth's surface with a force only 1/300,000 of that due to the Earth itself.

Weak though it may be, this force is the cause of the stability of celestial bodies; and because of it, the velocity needed for an object to escape from the Earth is 11,200 metres per second (roughly 7 miles per second). To escape from the Solar System, starting from a distance from the Sun equal to the radius of the Earth's orbit, an object would have to be given a starting velocity of over 43 kilometres per second. It is because of gravitation that the stars in our Galaxy make up a definite system.

The shape of the Galaxy is the result of equilibrium between gravitation and centrifugal force. Each star moves in the gravitational field of all the others; thus the Sun turns around the centre of the Galaxy with a velocity of about 250 kilometres per second, completing one revolution in roughly 250,000,000 years. To escape from the Galaxy, a star staying near the galactic plane would have to be given an initial velocity of at least 380 kilometres per second. Such relatively great velocities account for the flattened shape of the Galaxy. If a star-system does not rotate, it will not acquire a flattened shape. The spherical or nearly spherical systems which are known to exist near our Galaxy must have this shape either because they rotate very slowly, or because they do not rotate at all.

Consider next a star which does not lie near the galactic plane. It is obviously attracted by a force which is perpendicular to the plane, and directed toward the Galaxy. If its velocity in the direction perpendicular to the plane is small, it will not be able to move far from the plane; but if it has a high velocity in the direction perpendicular to the plane, it will be able to follow an orbit which will extend well outside the Galaxy. The differences in distribution of the stars in space are therefore of dynamical origin, but, as will be shown later these dynamical differences are associated with physical differences among the stars concerned.

Dwarfs and giants

The colour at the fire-hole of a furnace depends upon the interior temperature; it may be blood-red, cherry-red or white, in order of increasing temperature. In other words, the higher the temperature, the greater is the proportion of blue radiation compared with red, so that the proportion of short-wave to long-wave radiation increases with the temperature. Generally speaking, this is also true for the stars, whose colours are dependent upon their surface temperatures. A star's colour is measured by what is called its *colour index*, related to the ratio of blue to red radiation which is emitted. Red stars have a positive colour index; with blue stars, the colour index is negative.

Two essential facts must be known before a star can be put into its proper classification. One, already described, is the absolute magnitude; the other is the colour index. When the stars are plotted on a diagram according to their absolute magnitudes and colour index, it is at once clear that the stars are not scattered in a haphazard manner. Most of them lie near a line which crosses the diagram diagonally, and is called the Main Sequence.

A star cannot be plotted on the diagram unless its distance is known. A small number of nearby stars for which the parallax method is applicable can be very useful here, and the diagram for them is shown; but the stars of the Hyades, described earlier, are better known as a class, and, as noted, are used as reference standards in the calibration of the relationship between absolute magnitude and spectral type.

It has already been stated that the spectrum of a typical star is made up of a rainbow background crossed by dark absorption lines, each line being due to a particular element or group of elements. Another method for estimating stellar luminosities is that of studying the relative intensities of certain spectral lines, which, like the star-colours, depend upon surface temperature. In fact, there is a close relationship between the temperatures and the details of the spectra.

20 The light-curve of Delta Cephei. The magnitude range is from 3·6 to 5·1, with a period of 5·37 days. In other words, the star is 2·3 times brighter at maximum than at minimum. Cepheids are pulsating stars, which alternately expand and contract like a balloon. During their cycle of evolution, some stars pass through a 'variable' stage, which lasts for some thousands of millions of years. Cepheids are variable stars of this kind.

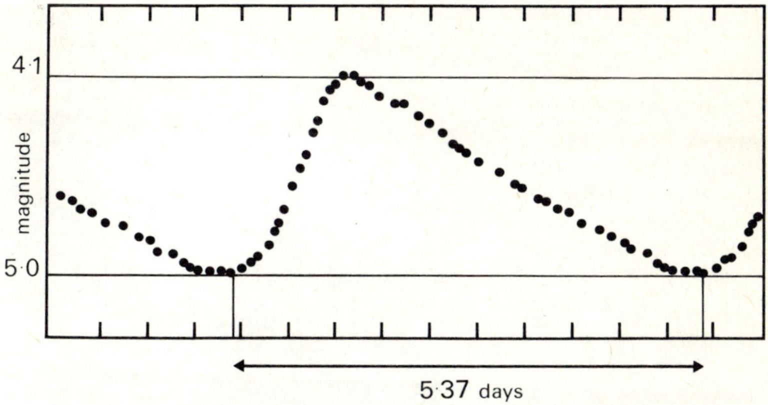

The initial discovery was that of a link between absolute magnitude and spectral type; this resulted in the well-known Hertzsprung-Russell or H/R diagram, named in honour of E.J. Hertzsprung of Denmark and H.N. Russell of the United States. It closely resembles the diagram shown in figure 28. The advantage of the colour index/absolute magnitude diagram is that it is more precise; classification of a star's spectrum is always less exact.

The spectral types of stars are given by different letters. Stars of type B are blue; type A stars are white, F and G yellow, and K and M red. It has already been noted that B-type dwarfs lie at a mean distance of about 150 light-years from the galactic plane, but it should be added that all normal B-stars are much hotter and more luminous than our Sun, so that the term 'dwarf' is somewhat misleading when applied to them. A dwarf star has a radius comparable with that of the Sun, and the radius depends upon the absolute magnitude to only a limited degree; thus a star with a luminosity 10,000 times that of the Sun may have a radius only 6 times as great, while a star with a luminosity of 1/10,000 that of the Sun will have a radius of one-third that of the Sun. Along the

21 The Pleiades photographed in colour from Palomar. Every astronomical enthusiast knows the 'Seven Sisters' but in fact the cluster contains many members, together with a nebula which shows up well only with long-exposure photography. The nebula is of the reflection type, that is, it shines only because it is lit up by the stars mixed up with it.

22 A diagram of the shape of the Galaxy. The stars, with the interstellar gas, form a vast flattened disk with a diameter of about 100,000 light-years. The Sun lies approximately in the main plane, at a distance of 30,000 light-years from the centre. Asterisks indicate the positions of globular clusters – huge collections of stars which move round the Galaxy.

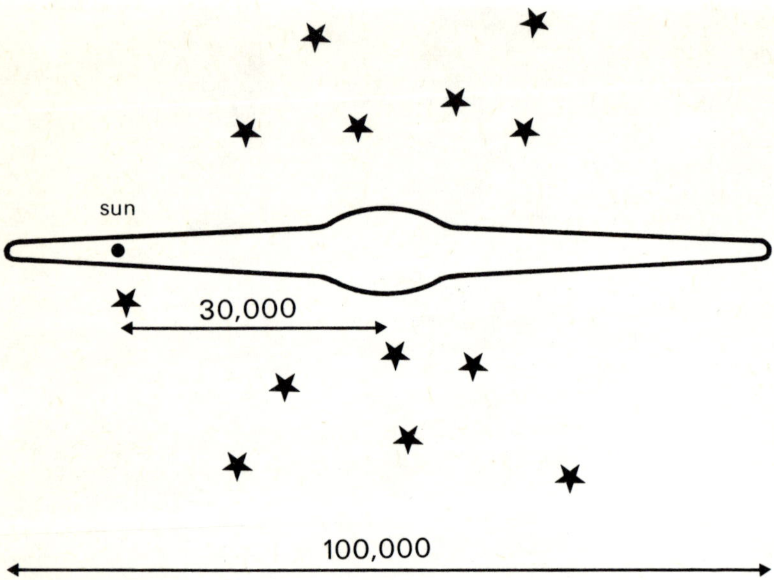

Main Sequence, we therefore find stars with a brilliancy range of 100,000,000 to 1, but a range in radius of only about 20 to 1. With two stars of the same colour index but different luminosity, the more brilliant star will have the larger radius.

Stars with radii much larger than those of the Main Sequence are known as *giants*. They are of immense size, and may have radii 10, 100 or even 1,000 times greater than the Sun's. The region in the H/R diagram containing stars of large positive colour index and high luminosity is the region of the red giant and supergiant stars. Red giants are found in large numbers in the globular clusters which lie near the boundary of the Galaxy, and in general there is a close relationship between the intrinsic properties of the stars and their motions, because the distribution of the stars in space is related to both. There is an important distinction between stars such as the red giants, well away from the galactic plane, and ordinary stars of the Main Sequence, which are in general scattered around the neighbourhood of the plane.

Spectroscopically, giants are easily distinguished from dwarfs because the absorption lines are much thinner. When the spectrum of a star can be examined in sufficient detail, there is little difficulty in working out its type, its colour, its place in the H/R diagram and its position in the colour index/absolute magnitude diagram. In many cases this is the only known means of determining the absolute magnitude of a star, and hence its distance. Distances worked out by this method of *spectroscopic parallax* are naturally rather uncertain, because one can never be sure that the derived luminosities are wholly reliable, but in general the method is a satisfactory replacement for ordinary or *trigonometrical* parallax measurement. However, care must be taken to avoid reasoning in a circle. The Hertzsprung-Russell diagram was compiled by using stars of known distance, but the diagram itself is used to measure the distances of stars for which other methods fail.

This method of spectroscopic parallax has been of fundamental importance in fixing the positions of the stars in space. For thousands of remote stars, no other means can be used. It is in this way, then, that it has been possible to classify the stars into different types, according to their distances from the main plane of the Galaxy. Essentially, the stars close to the galactic plane are of what is termed Population I, while stars well away from the plane belong to Population II. Generally speaking, Main Sequence stars are of Population I, red giants of Population II; it is also true to say that young stars belong to Population I, old stars to Population II.

Gas and dust

Great masses of dust and gas can be seen in the Galaxy. The gas is visible because it is luminous, while the dust absorbs the light of more remote stars and so betrays its presence. For instance, it is

23 The solar dome at the Crimean Astrophysical Observatory, USSR. The main equipment, designed by Dr Severny, is contained in this dome. (Photographed in 1960).

well known that in the southern hemisphere of the sky, the Milky Way apparently splits into two branches. However, this division is not real; it is due to the presence of a large, dense dust-cloud relatively close to the Solar System (close, that is to say, on the cosmical scale!) which absorbs the light from the stars behind it. Dust is very efficient at absorbing light, and in the plane of the Galaxy it causes the apparent magnitude of a star to increase by 2 for every 3,000 light-years; thus a star 6,000 light-years away from us will appear two magnitudes fainter than it would otherwise do, while a star in the centre of the Galaxy will have its magnitude increased by 20 because of the dust absorption. A further 15 magnitudes must be added on account of the star's distance, and so the absolute magnitude must be increased by 35 to obtain the apparent magnitude that the star would have if it could be observed at the centre of the Galaxy. Since stars of magnitude greater than +23 are too faint to be photographed, a star lying at the galactic centre would have to be of absolute magnitude −12 or more to be observable. No star as luminous as this can exist, and it is evident that the dust hides the centre of the Galaxy completely so far as we are concerned.

The bright nebulæ account for only a small fraction of the total mass of gas in the Galaxy. Generally speaking, the nebulæ shine because they are affected by ultra-violet radiation sent out by hot stars close to them. This ultra-violet radiation acts on the atoms making up the gas, and excites the atoms to a level at which they emit light. Analysis of the light makes it possible to find out the chemical composition of the nebulæ.

Visual observations of the absorbing gases and the luminous nebulæ show that the gas and dust clouds are spread out close to the galactic plane. As will be shown later in this book, there is nothing fortuitous about the similarity between the distribution of the gas and dust, and the distribution of most of the stars belonging to the Main Sequence.

Recently-developed techniques have led to important discoveries about the Galaxy. Consider, first, the classic diagram which shows

24 Part of the Milky Way in the southern hemisphere. Note the dark obscuring mass of gas close to the two brightest stars.

the so-called *electromagnetic spectrum*, or total range of wavelengths, ranging from very short wavelength X-rays up to long-wavelength radio waves. Extremely short wavelengths are measured in Ångströms or microns; one Ångström is equal to one ten-millionth of a millimetre, while one micron is equal to one-thousandth of a millimetre. X-rays may have wavelengths of no more than 1 Ångström, which is small by any standard. The human eye is sensitive only to the narrow band in the electromagnetic spectrum between 0·4 and 0·8 microns. The photographic plate extends this range by about 2 microns toward the longer-wave end, and in theory it should provide a greater extension toward the ultra-violet, but unfortunately this extension is theoretical only, because all the ultra-violet and X-rays coming from the sky are blocked out by the Earth's atmosphere. The situation is a little better on the infra-red side of the 'optical window', where wavelengths in the millimetre and decametre range can pass through the atmosphere more or less unobstructed; but still longer radiations, with wavelengths in the kilometre range, are completely blocked. However, the 13 or 14 octaves accessible in the 'radio window' have led to a great deal of extra information about the Galaxy. Observations from ground level have been very successful, thanks to the new equipment which has been developed during the past twenty years. On the other hand, the short-wavelength radiations cannot be studied by means of instruments set up on the Earth's surface, and there is no choice but to send up the equipment in rockets, artificial satellites, or (for convenience) high-altitude balloons.

Radio radiation (lines)

Radio emission provides a continuous background emission together with a line spectrum. The continuous emission is of an intensity which changes slowly with wavelength, and there are very few definite spectral lines. At the present moment one strong emission line has been detected; it is caused by atoms of hydrogen, and has a wavelength of 21 centimetres. Several weak lines of hydrogen have also been detected. There is also an absorption line at 18 centimetres, due to molecules of OH (oxygen combined with hydrogen). These lines were discovered by observation after their existence had been predicted by physicists. The 21-centimetre hydrogen line has proved to be of particular importance, both in atomic physics and in studies of the structure of the Galaxy.

The discrete emission or absorption of radiation corresponds to a change in the energy-state of an atom. The atom of hydrogen, for example, consists of one proton and one electron; the transition which produces the 21-centimetre line is caused by a change in the energy level affecting both particles. Both the proton and the electron have axial rotation, and this gives rise to measurable magnetic effects. According to whether the magnetic moments of the electron and the proton are in the same or in the opposite direction, the energy of the hydrogen atom will be slightly different, and this admittedly very slight difference between the two states of the hydrogen atom accounts for the 21-centimetre line. The mechanism involved is basically very simple. Atoms of hydrogen in interstellar space collide, with increase of energy; the return to a lower energy-level is accompanied by the emission of radiation at a wavelength of 21 centimetres. The study of this line leads to a very accurate knowledge of the distribution of hydrogen atoms in the Galaxy, and maps of the hydrogen clouds can be drawn up. The density of the hydrogen atoms is, at maximum, 2 or 3 per cubic centimetre; the denser clouds stretch through the Galaxy rather in the manner of huge arms. Visual and radio researches agree in showing that the hydrogen clouds are spread out in the neighbourhood of the galactic plane.

The structure of the Galaxy has been satisfactorily studied for the region lying at more than 10,000 light-years from the galactic centre. However, the 21-centimetre investigations have shown that totally different conditions occur near the actual centre. Inside a sphere with radius about 600 light-years there is a mass of hydrogen in a state of violent agitation, with velocities reaching 200

25 The planetary nebula NGC 7293 in Aquarius, photographed with the 48-in Schmidt telescope at Palomar. This is a relatively bright planetary, comparable with the Ring Nebula in Lyra; it is also the same kind, though less symmetrical. For some reason it was not included in the original catalogue of nebulous objects drawn up in the eighteenth century by Messier.

26 The electromagnetic spectrum. The band below the diagram shows the extent of atmospheric absorption for the various wavelengths. Above 25 km the atmosphere is transparent for wavelengths shorter than 20 Å.

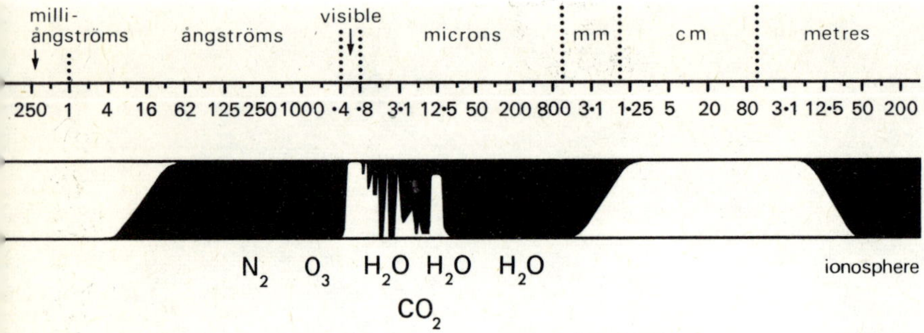

kilometers per second; there is also a general expansion at a rate of about 50 kilometres per second. The quantity of hydrogen in this central region has about 30 or 40 million times the mass of the Sun, and its expansion corresponds to a flux of material of about one solar mass per year.

Radio radiation (continuous emission)

Two kinds of continuous emission have been identified, and have been named *thermal* and *non-thermal.* Thermal emission results from encounters between electrons and ions (an ion being part of an atom that has been broken up). Electrons are of relatively low mass, and are negatively charged. When they move close to a more massive ion which is positively charged, the electrons will undergo a 'braking' effect, and will emit radiation. So long as the electrons owe their velocities to thermal agitation, their emissions will be of a characteristic kind, and the resulting radiation will depend upon the properties of the electrons which are being affected by the thermal agitation. On the other hand, the character of the emission changes completely as soon as the motions of the electrons become in any way regular, and in this connection the effects of magnetic fields are of vital importance so far as radio astronomy is concerned. When moving in a magnetic field, an electron turns around

27 The two states of the hydrogen atom concerned with the emission of the 21-cm line. To the left, the magnetic moment of the proton and the electron are parallel and in the same sense; to the right, the sense is opposite. During transition from one state to the other, a certain amount of energy is emitted in the form of electromagnetic radiation and this produces the 21-cm line.

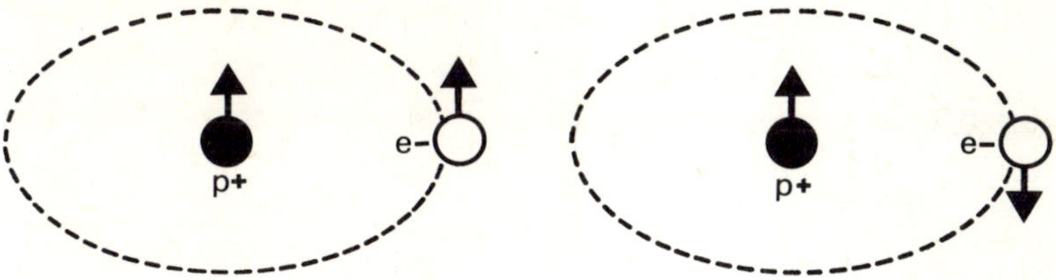

the lines of force, and the centrifugal force is compensated by the so-called Laplace force (that is to say, the action of a magnetic field upon a moving charge). The electron suffers acceleration, and emits radiation. In this case, obviously, the movement of the electron is being controlled by the magnetic field, so that its radiation is easily distinguishable from that due to braking in the vicinity of a positively-charged particle. The radio emission caused by spiral movement in a magnetic field is very intense, with high-energy electrons, and is comparable with the energy of cosmic-ray particles.

This new type of radiation was detected by the use of machines in which high-energy electrons are forced to circulate in a magnetic field. Its official name, *synchrotron radiation*, is taken from the name of the machine in which it was originally found.

Synchrotron radiation is therefore very different from thermal radiation. The latter is characterised by the fact that its intensity falls off rapidly both for very long and for very short wavelengths. The non-thermal radiation of high-energy electrons is quite different; the intensity decreases in a very regular manner from radio wavelengths through to the visual band and beyond, perhaps as far as the X-ray region.

Both types of radiation have been detected in the Galaxy. Thermal emission is associated with the presence of ionised hydrogen, and is observed chiefly in the decametre band. It shows that

28 A colour-magnitude diagram for the Pleiades cluster. The Main Sequence is clearly shown. The diagram is printed in colour to emphasise the dominance of blue to the left of the diagram and the dominance of red to the right.

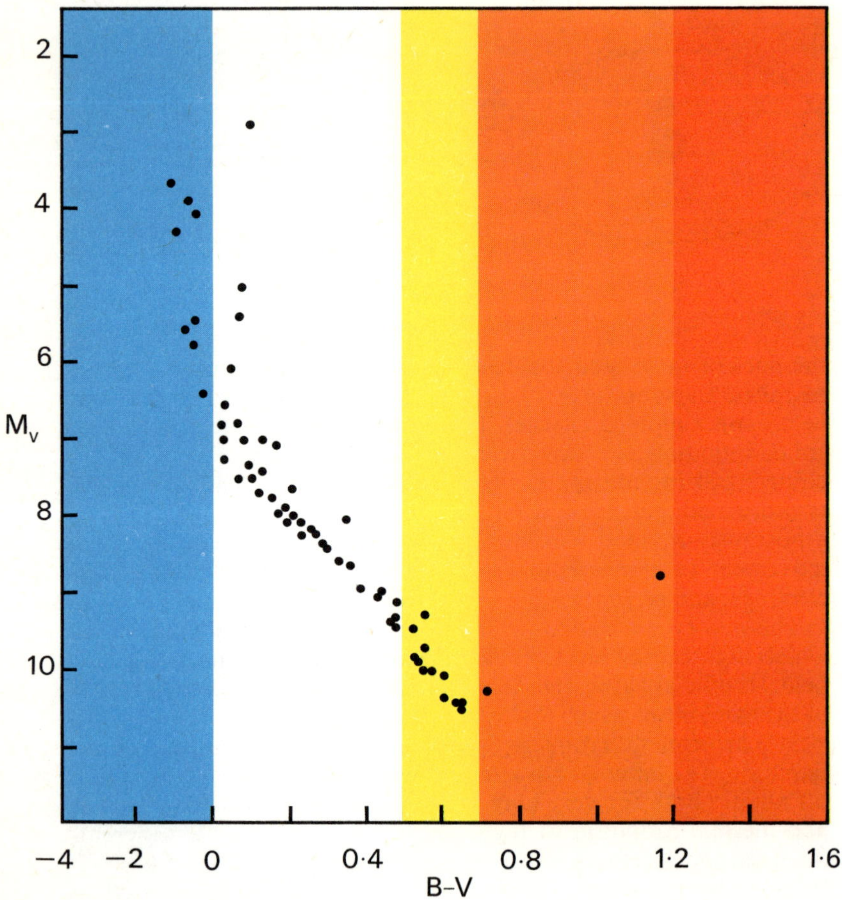

the distribution of the gas responsible for the emissions is much the same as the general distribution of gas throughout the Galaxy. However, the non-thermal emission is associated with individual objects, the discrete radio sources, formerly (and most misleadingly) known as radio stars. Some of the discrete sources lie outside our Galaxy, and will be discussed in part 5. Others are contained inside the Galaxy; of these, one of the most important is the Crab Nebula, a mass of gas in the constellation of Taurus (the Bull). It lies at a distance of about 4,000 light-years, and is almost 5 light-years in diameter; it has an emission spectrum in the optical range, in which lines of hydrogen, oxygen and nitrogen are prominent, and it also shows a non-thermal radio spectrum. The Crab Nebula is a remarkable object, and is known to be the remnant of a supernova. In 1054, Chinese astronomers observed a new star which became so bright that it could be seen in broad daylight, and remained visible for several months before fading away. The present-day Crab Nebula is all that remains of this tremendous explosion; not surprisingly, the spectrum is of a very unusual type.

As well as these discrete sources, weak non-thermal radio radiation occurs almost everywhere in the Galaxy, and gives evidence of the existence of high-energy electrons and magnetic fields in the space between the stars. The central region of the Galaxy contains a source of thermal emission with a diameter of about 30 light-years, together with a non-thermal source of about the same size.

X-rays

X-ray detectors can be flown in high-altitude rockets, so providing material for a map of the X-ray emission from the sky. Various X-ray sources have been detected in this way, of which one of the strongest is the remarkable Crab Nebula. Therefore, the spectrum of the Crab extends from the radio range right through the optical band through to the very short X-radiations. An even stronger X-ray source has been located in the constellation of Scorpio.

Table 1 Radius of gyration of protons moving in a magnetic field of strength one-millionth of a gauss

Energy (electron volts)	Radius of gyration
1	1,400 kilometres
1,000	45,000 kilometres
1,000,000	1,400,000 kilometres
10^{9}	0·38 astronomical units*
10^{12}	220 astronomical units
10^{15}	3·5 light-years
10^{18}	3,500 light-years
10^{21}	3,500,000 light-years

* One astronomical unit is the distance between the Earth and the Sun; 150,000,000 kilometres = 93,000,000 miles = 500 light-seconds.
10^{-6} gauss is the order of magnitude of the galactic-magnetic field.

Cosmic rays

Studies of our Galaxy and external systems cannot be separated from researches into cosmic rays, which are not true rays at all, but are high-velocity atomic particles (mainly protons) with energies greater than 1,000 million electron volts. The total flux of the particles which reach the Earth with energies of over 5,000 million electron volts is about 0·2 particles per square centimetre per second. The distribution of the energy of these particles shows a very regular decrease; the flux falls off as the 1·7 power of the energy, and this decrease continues up to very high energy levels. Particles with energies as high as 10^{20} electron volts have been detected. As will be shown below, weak magnetic fields seem to be present everywhere in the Galaxy, and we must next discuss the motions of charged particles in a magnetic field.

29 The radio interferometer at the Mullard Radio Astronomy Observatory in Cambridge. Not all the equipment can be shown in one photograph, for it covers a considerable area. With an interferometer of this sort, the greater the area the greater the resolving power. The interferometer is not, of course, steerable in the same way as the Jodrell Bank paraboloid, but in its own way it is equally valuable. (Photographed in 1964).

Movements of particles in a magnetic field

Electrically charged particles move around the lines of magnetic force, so that the centrifugal force is balanced by the Laplace force. The radius of movement depends upon the charge, the mass and the velocity of the particle concerned, together with the strength of the magnetic field. For a proton moving in a magnetic field of 10^{-6} gauss, the radius of gyration is summarised in table 1. The gauss is the unit of intensity of a magnetic field; the magnetic induction at a given point is 1 gauss when the maximum electromotive force that can be induced in a conductor 1 cm long moving through the point with a velocity of 1 centimetre per second is one unit of electromotive force.

Clearly, particles with energies of the order of 10^9 electron volts cannot escape from the Galaxy, because their radii of gyration are

30 The Trifid Nebula, M20 in Sagittarius, photographed with the 200-in Hale reflector at Palomar. This is a magnificent example of its type, and even a small telescope will show it, but as in most other cases, long-exposure photography is needed to show the fine structure. Moreover, the Nebula lies in the southern part of the sky, and is always inconveniently low when viewed from Britain and most of the United States.

too small, but particles with energies of the order of 10^{18} electron volts cannot stay permanently in our Galaxy. Since we must also take non-uniform magnetic fields into consideration, it is found that particles of energy well below this value can also escape under suitable conditions.

Charged particles move in spiral paths when in a uniform and constant magnetic field. If the field is constant but subject to slow variation from one region of space to another, the movement of the particle becomes more complicated. When it moves toward a region of higher intensity of the magnetic field, it will travel parallel to the lines of force, slowing down as it does so. This slackening in velocity will continue until the moment when the direction of travel is reversed. Regions where the magnetic field is strong enough to cause this effect act as reflecting media, and are often called *magnetic mirrors*.

In a magnetic field which changes slowly with time – slowly, that is to say, compared with the time needed by the particle to complete its journey round the field – a particle will be affected by a weak inductive magnetic field which will modify its path, making it travel across the lines of force.

The phenomenon of reflection at regions where the magnetic field is sufficiently intense, together with the effects of motion across the lines of force, greatly complicates the movements of all charged particles which encounter intense or variable magnetic fields. This complicated random motion of a large number of particles is what is termed the phenomenon of *diffusion*.

Under these conditions, the movements of particles are completely random in direction. If a group of particles starts off from

any given point, they will scatter in all possible directions. For the sake of clarity, suppose that the particles are able to move only by units of length l in any direction. After N displacements, the arrival points of the particles will be scattered over the inner surface of a sphere of radius $l\sqrt{N}$. If the velocity is v, then each movement will need a time $(l/v)=t_0$, and N movements will be completed in a time Nt_0. The distance R will be covered after $(R/l)^2=N$ movements, taking up a time (R^2/lv). With cosmic ray particles, the distances l are of the order of 30 light-years, and the velocities approach the velocity of light. Cosmic rays would take something like a thousand million years to diffuse through a sphere of radius 50,000 light-years. This means that the cosmic ray particles cannot be permanent members of our Galaxy, and there must be some process whereby they are constantly renewed. This problem will be discussed later.

The magnetic field of the Galaxy

The strength of the general magnetic field of the Galaxy may be estimated in several ways. The oldest and best of these concerns the polarisation of starlight, which is important enough to be discussed here in some detail.

The polarisation of light

Light has properties both of wave-motions and of corpuscles. The classic work of Clerk Maxwell in the nineteenth century showed that the wave properties of light can be well represented by the propagation of an electric field and a magnetic field, perpendicular to each other and perpendicular to the direction of the light-ray. At a fixed point, the electric field and the magnetic field oscillate at the frequency of the radiation; it is, of course, the electric field which acts on suitable receivers, such as the human eye or the sensitive plate. The direction in which the electric field vibrates is one of the characteristics of the radiation. If the electric vibrations are orientated – that is, constantly aligned along one particular direction instead of being distributed at random – the light is polarised. Ordinary light is the result of the superposition of large numbers of waves vibrating in random directions. When one direction of the electric field is of higher intensity than the others, the light is partially polarised; this too is observed in nature.

It has already been noted that light can be emitted because of the movements of electrons. When these movements are due simply to thermal agitation, the light is non-polarised, and is called thermal

31 A diagrammatic representation of the motion of a charged particle in a non-uniform magnetic field.

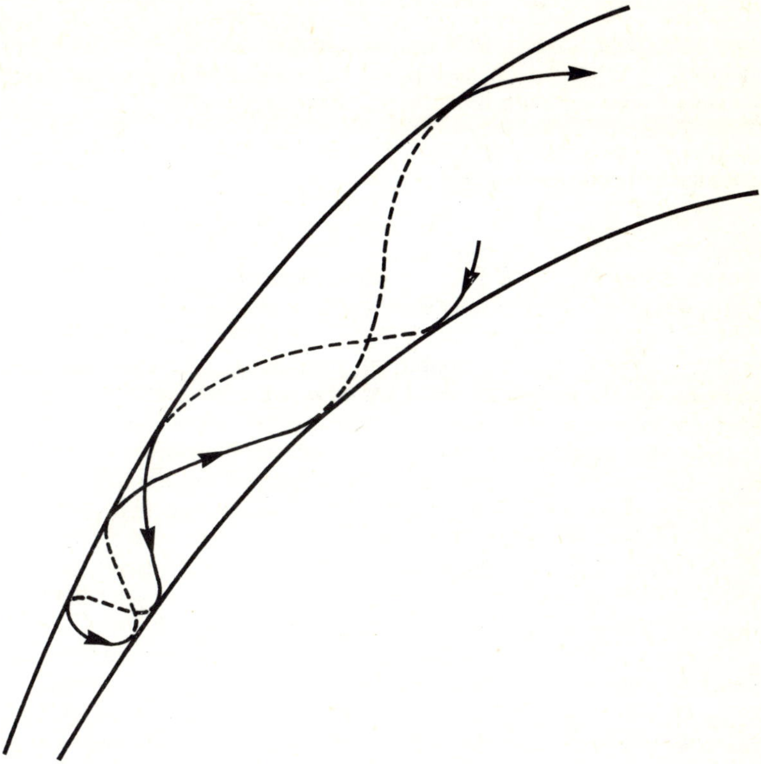

radiation. But radiation due to the spiralling of electrons in a magnetic field comes from an anisotropic source and is strongly polarised. Polarisation can also be produced when radiation crosses a medium with anisotropic characteristics.

Interstellar polarisation

In 1948 it was found that starlight is slightly polarised. This phenomenon is rather surprising, because there seems no reason why a star, which is a symmetrical sphere, should emit polarised radiation. The only possible solution is to assume that the polarisation is due to the journey made by the light-rays across interstellar space. The only force capable of producing effects of this sort appears to be a magnetic field; yet how can a magnetic field act in such a way?

The absorption of light in space is due to the presence there of fine dust, and the anisotropic properties must presumably be the result of the action of the magnetic field upon this dust. Nothing definite is known about the chemical composition of the dust, but it is certain that atoms of iron, oxygen and silicon, together with various types of molecules, exist between the stars, so that the dust is presumably a conglomerate of indefinite composition, which comes together because of the random encounters between different kinds of atoms and molecules. Under such conditions it would be thought that the dust particles should be spherical, in which case there would be no orientation in a magnetic field; it is not easy to see how such conglomerates could have developed into needle-like or plate-like forms. On the other hand, various chemically pure bodies have different properties. Laboratory experiments have shown that there are at least two examples of small crystals with very marked anistropy, which develop naturally; in a rarefied atmosphere, atoms of iron can collect to form needles, while carbon condenses naturally into plates of graphite. Therefore, it is possible that part of the interstellar dust is made up of microcrystals of graphite, which would be orientated by the magnetic

32 The Horse's-Head Nebula, NGC 2024 in Orion, near the star Zeta Orionis, photographed with the 48-in Schmidt telescope at Palomar. The dark nebula has a slight resemblance to a knight in chess – hence the nickname. There is no difference between a luminous and a dark nebula except that in a dark nebula there are no stars suitably placed either to light it up or to make it luminous.

Table 2

Rotation of the plane of polarisation (in degrees).
Magnetic field of 10^{-6} gauss.
Electron density of 0·1 electron per cubic centimetre.
Path-length of 3·26 light-years (= 1 parsec).
Wavelength (in metres).

	0·1	**1**	**10**
Rotation	$0^{\circ}{\cdot}465$	$46^{\circ}{\cdot}5$	4650°

field. Under these conditions, polarisation would occur.

The particles must have dimensions of the order of one micron (one-thousandth of a millimetre). After the journey across interstellar space, the electric field of the radiation parallel to the magnetic field has become very weak, so that the interstellar dust is responsible for partial polarisation. Measurements of the polarisation produced indicate that the strength of the general magnetic field in space is about one-millionth of a gauss.

Synchrotron radiation

It has already been noted that there are important spectral differences between radio waves of thermal origin and synchrotron radiation. This means that the two types of radiation can be distinguished in the general emission coming from the Galaxy. With wavelengths of the order of 10 cm, synchrotron radiation is dominant; given the numbers and energies of the electrons which emit the galactic synchrotron radiation, it is possible to deduce the strength of the general magnetic field of the Galaxy. Unfortunately, the data for the numbers and energies of the electrons are not known accurately, and various assumptions have to be made. Since high-energy electrons are responsible for the synchrotron radiation, it is logical to suppose that their characteristics are much the same as for cosmic ray particles (or, more properly, the electronic component of cosmic radiation). If so, the strength of the interstellar magnetic field must be approximately 1×10^{-6} gauss.

Faraday rotation

The propagation of electromagnetic radiation in the Galaxy is affected by the general galactic magnetic field. Consider, for instance, a polarised electromagnetic wave with a propagation along the direction of the magnetic field. A linearly-polarised wave has its own electric field, which, at any set point in space, vibrates in a plane agreeing with its direction of motion; that is to say, its plane of polarisation. During its movement, however, it is affected by the spiralling of the electrons in the magnetic field, and the plane of polarisation will rotate progressively, causing what is termed the Faraday effect. Obviously, the Faraday effect will depend upon the strength of the magnetic field, the electron density, and the distance covered by the wave.

Table 2 gives the values for the rotation of the plane of polarisation, expressed in degrees, for a magnetic field of 10^{-6} gauss, an electron density of 0·1 electron per cubic centimetre, and a path-length of 3·26 light-years. (This distance – 3·26 light-years – is also known technically as one *parsec*.)

Measurements of the directional differences of the planes of polarisation for different wavelengths, from extragalactic radio sources, give some idea of the strength of the magnetic field of the Galaxy, provided that the electron density is known. These measures indicate that the magnetic field is greater than 2×10^{-6} gauss.

Radio sources and cosmic radiation

It has been shown that most of the non-thermal radio radiation from the Galaxy comes from discrete sources, and that it is due to the movement of high-energy electrons in a magnetic field; there are various means of estimating the strength of the galactic magnetic field, and it has also been noted that cosmic radiation is spread all through the Galaxy.

The presence of high-energy electrons in galactic radio sources, such as the Crab Nebula, indicates that these sources contain protons and other high-energy nuclei, and that the sources are responsible for sending out the cosmic ray particles. In other words, there is a close connection between cosmic rays and non-thermal radio sources. The mechanism of cosmic-ray production will therefore be better understood when the conditions inside the non-thermal radio sources have been determined.

In the Crab Nebula, for instance, the number of high-energy electrons must be about 1 per litre; their energy is of the order of 10^{12} electron volts, and they radiate in a magnetic field of about 1×10^{-3} gauss. Such electrons must lose all their energy in a few decades; the energy will have been radiated away. Yet the Crab Nebula has certainly existed for more than 900 years (remember that the supernova responsible for it was watched as long ago as the year 1054). High-energy electrons must be continuously replenished, and so the appropriate energy-sources must be active all the time. Such energy-sources are easily found. High-energy protons collide with slow electrons; these electrons are speeded up, and it is the accelerated, now fast-moving electrons which are responsible for the non-thermal radiation.

There are great differences between protons and electrons. The high-energy protons can last for much longer than the electrons because they lose no energy through radiation and very little by collision. The energy spectra of the electrons produced by collisions with the protons of high energy is a faithful representation of the energy spectra of the protons themselves.

It might therefore be supposed that all supernova explosions are likely to produce cosmic radiation. If so, then from time to time (perhaps once per century) a fresh supernova outburst will inject a further supply of cosmic ray particles into the Galaxy. Part of this

33 The Crab Nebula, M1, photographed in colour with the 200-in. Hale reflector at Palomar. This Nebula is the remnant of the supernova observed in 1054 by Chinese and Japanese astronomers. Its distance is 4,000 light-years and the gases are still expanding outwards from the explosion-centre. The Crab Nebula is a source of radio emission and X-radiation and is one of the most significant objects in the Galaxy. Nothing else quite like it is known.

radiation would escape into intergalactic space; braking would account for the loss of another part, and the rest would receive extra acceleration within the Galaxy. The composition, intensity and energy of cosmic radiation, as observed from the Earth, seem to be in substantial agreement with this idea.

Conclusion

Following a description of the observational methods and results, it may be useful to draw up a balance-sheet, if only to put all these ideas into their proper perspective. First there is the question of distance – and here, an everyday comparison will be of help. A ten-centimes coin placed under the Arc de Triomphe, in Paris, will subtend an angle of 1 second of arc as seen from the Carrousel (Tuileries). The ratio between the diameter of the coin and the distance between the Arc de Triomphe and the Carrousel is very nearly the same as the ratio between the radius of the Earth's orbit round the Sun, and the distance of Proxima Centauri, the nearest star. The Sun is 11,000 terrestrial diameters away; the nearest star is 250,000 times more distant than the Sun; the distance from the Sun to the centre of the Galaxy is about 10,000 times the distance between ourselves and Proxima Centauri.

Much depends upon what is meant by the terms 'near' and 'far'. Inside the Solar System, 'near the Earth' means a few million kilometres; with respect to the stars, 'near the Sun' means a few light-years. Stars which are some thousands of light-years away are said to be 'distant', but, as will be described later, a 'near' galaxy is situated at a distance of millions of light-years.

There is also the question of the time-lag between the moment when the light is emitted by a star and the moment when we see it. For instance, a nova 2,000 light-years away will have suffered its actual outburst 2,000 years before we see it. From a scientific point of view, however, this does not matter in the least; there is no need to know the exact date of any particular event of this kind. With individual stars, it is the chronological sequence of events which is important, and whether the events began 10, 100 or 1,000 years ago is of little consequence. With large groups of stars, statistical methods take over, and any effects due to the time-lag disappear. Moreover, the delays are insignificant with respect to the evolution of a star or galaxy. It is quite true that there are some dynamical effects, because the stars are never where we see them; they have moved on in their paths since they emitted the light now reaching us, and yet, taken as a whole, the Galaxy is evolving so slowly that effects of this kind are really not important.

The Earth moves round the Sun, and the radius of its orbit is the basis of all distance-determinations. The Sun itself is a star with no particular noteworthy characteristics; it moves in a somewhat eccentric orbit round the centre of the Galaxy, taking 250,000,000 years to complete one journey. At present its distance from the galactic centre is about 30,000 light-years.

Around us are the other stars of the Galaxy, and between them there lies nebulous material composed of gas and dust. It is known that one-third of the galactic mass is contained in the visible stars, while one-sixth of it appears as nebulous material. There remains a mass equal to about one-half of the total, about which nothing certain is known.

There are about 100,000 million stars in the Galaxy; they may be divided into two different 'populations', and the movements of these two groups may be distinguished. Also, the distribution of the stars in the Galaxy is associated with their types of movements. The Galaxy is flattened in shape, because it is in a state of rotation, and because of the way in which the different groups of stars are spread out above and below the main plane of the system. And just as the two definite stellar populations differ in movement, so also the stars in them differ physically.

The galactic gas, like the stars, is made up essentially of hydrogen, which is the principal constituent of the universe. This interstellar gas carries with it magnetic fields, which may be detected because they cause the polarisation of starlight and the rotation of the plane of polarisation of radio waves. The magnetic

34 Positions of the X-ray sources detected by Bowyer, Chubb and Friedmann by means of detectors flown in rockets. The numbers indicate the probable identities: 1 Crab Nebula; 2, 3, and 4 sources in Scorpio; 5 Kepler's supernova in Ophiuchus, seen in the year 1604; 6 galactic centre in Sagittarius; 7 source in Sagittarius; 8 source in Serpens; 9 source in Cygnus.

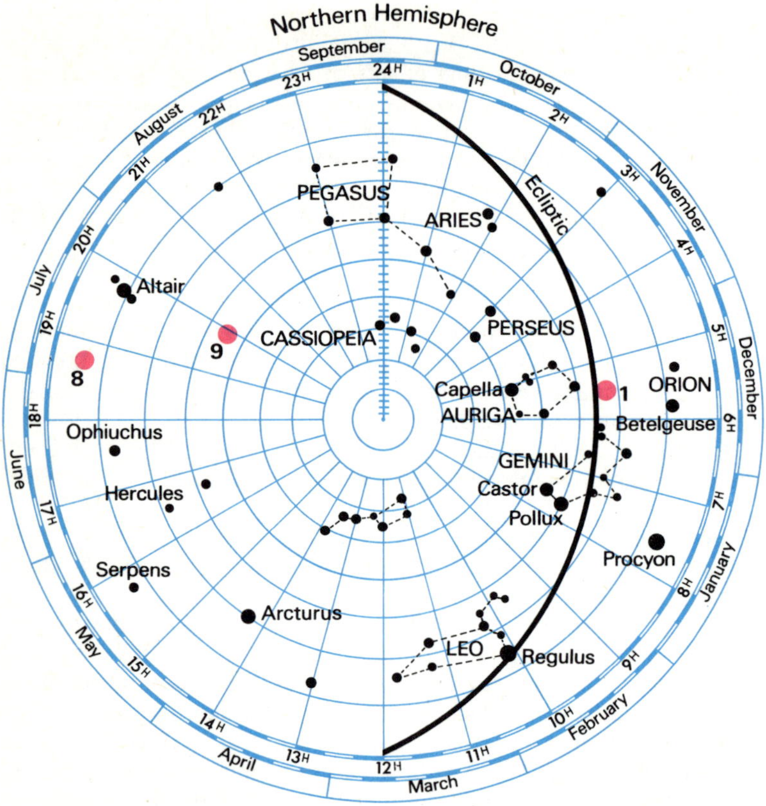

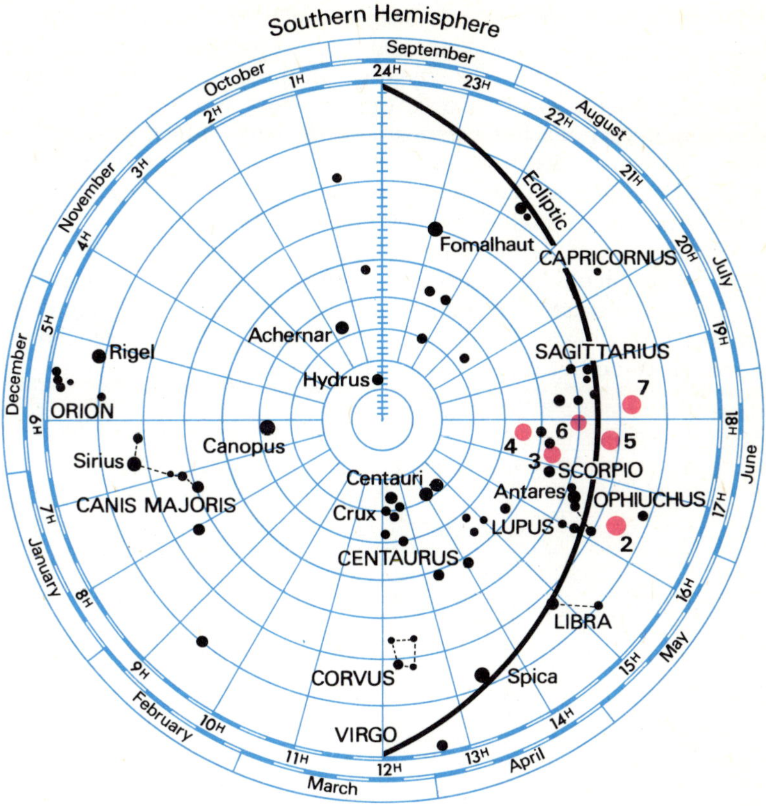
Southern Hemisphere
September
October
November
December
January
February
March
April
May
June
July
August
24H
23H
22H
21H
20H
19H
18H
17H
16H
15H
14H
13H
12H
11H
10H
9H
8H
7H
6H
5H
4H
3H
2H
1H
Ecliptic
Fomalhaut
CAPRICORNUS
Achernar
Rigel
Hydrus
SAGITTARIUS
ORION
7
6
4
5
3
Canopus
Sirius
SCORPIO
Centauri
Antares
CANIS MAJORIS
OPHIUCHUS
Crux
LUPUS
2
CENTAURUS
LIBRA
CORVUS
Spica
VIRGO

field has a structure which is both complex and turbulent, and it controls the movements of the particles of galactic radiation. The gas itself has an average density of about three hydrogen atoms per cubic centimetre, which works out at about one solar mass per cubic light-year. To this we must add the stars, which account for about two solar masses per cubic light-year. About one particle out of 10,000 million is a cosmic-ray particle, and there is one grain of dust for every 10,000,000,000,000 atoms of hydrogen.

Of course, astronomers are familiar with all these numbers, and have little difficulty in understanding their significance. To draw up an accurate picture of the Galaxy as a whole, and to understand the relative crowding of what is so often called 'empty space', it is essential to look at things from a distance.

3 Time

The universe evolves in time, and therefore we need a chronology which must be as exact as possible, bearing in mind that we have to deal with periods of tremendous length.

Time in the mechanical sense

Studies of the motions of bodies are based upon relationships between time, the forces concerned, and the amount of space involved. To begin with, the principle of inertia, laid down originally by Galileo, states that a moving body will cover equal distances in equal times, in the absence of any external applied force. Of course, this state of affairs – the absence of external force – is never encountered in the universe. It represents an ideal of classical mechanics, but it can never occur. However, the rotation of the Earth on its axis provides a fairly close approximation. If the Earth were a perfectly rigid sphere, all the forces acting upon it would be exactly balanced; its rotation, due solely to inertia, would be absolutely uniform, and it would rotate through 1/86,400 of its circumference in a period of one second of time.

This, needless to say, does not happen. Twenty years ago it was found that the length of the 'day' is not constant, and in 1950 astronomers agreed to define the 'second' as a fraction of the year 1900. It was assumed that 1900 lasted for 31,556,925·9747 seconds, and this value is regarded as reliable, so that it serves as a standard for the future. Yet we cannot be sure that this is precisely the same as mechanical time, and so it has been given the name of Ephemeris Time.

Having defined the second as a definite fraction of the year, the notions of acceleration and force follow. Acceleration expresses the change in velocity with time. With bodies falling freely, the velocity increases by equal amounts in equal time-intervals, and this is, of course, the acceleration. At the Earth's surface, the acceleration due to gravity is 9·81 metres per second, so that in free fall a body will increase its velocity by 9·81 metres per second in each second.

The acceleration of movement is due to a force which acts on the body. This force is proportional to the acceleration, and the constant of proportionality is the mass – usually known as the *inertial mass*, since it is the inertia of the mass which acts in opposition to the motion. With uniform circular motion, the velocity changes direction continuously, because it must remain tangential to the circle. An instantaneous change in direction will therefore cause an acceleration toward the centre of the circle, often called the centrifugal acceleration. With a planet moving round the Sun, the acceleration due to its orbital motion is equal to the acceleration caused by the Sun's attraction, so that the attractive force is proportional to the *weight mass*, that is, the mass as attracted by the Sun. In classical mechanics, as in relativity, inertial mass and weight mass are regarded as identical.

Time in the mechanical sense depends both upon acceleration and upon the force of attraction. This is easily seen by considering the behaviour of a pendulum clock, in which the swing of the pendulum is controlled by the Earth's gravitational pull; this force must therefore be taken into account when defining the period of the swing.

Gravitational forces are responsible for the motions of all the bodies in the universe, and the equivalence of gravity and acceleration is clearly shown. And since 'time' must be included in any definition of acceleration, the verification of the laws of motion means that mechanical time must be the same for the movement of the Earth round the Sun, the movements of stars in our Galaxy, and for the movements of stars in other galaxies.

Time and energy

There are various forms of energy, such as the kinetic energy due to motion, potential energy in a gravitational field, radiant energy, the energy due to thermal agitation of molecules of a gas, and the internal energy of an atom. Experience proves that these different forms of energy can be transformed from one to another. The law of the conservation of energy is fundamental in physics.

The transformation of energy is never instantaneous; it takes a certain period of time. For any system, it is possible to define the quantity of energy which is transformed, in unit time, into another energy-form. The quantity of energy per unit of time is known as the power. Horse-power is a very common unit; the official unit is the watt. In a hydro-electric installation, for example, energy is in reserve behind the dam in the form of potential energy; in the flow pipes, it is changed into kinetic energy; and when the kinetic energy of the moving water turns the rotors of the dynamos, the result is a change into electrical energy. The quantity of electrical energy which enters the grid per second is equal to the power available in each second of time at the highest water-level in the system locked up in the form of potential energy – taking into account, of course, the inevitable losses due to such causes as rises in the temperature of the water, kinetic energy of the dynamo rotors, and the energy wasted upon the low-level water flowing out of the system.

When the capacity of the system is known, together with the amount of potential energy held in reserve behind the dam, it is not difficult to calculate the period during which the system can be supplied without the need for providing extra water behind the dam. Since energy is the produce of power by time, it is clear that each release of energy must take a certain time, and the relationship between time, energy and power may be considered as a novel way of defining 'time'.

Time and light

Light is a vibratory phenomenon, and the unit of time can therefore be defined as the period necessary to complete a set number of vibrations. Visible light presents difficulties, because the frequency is so high that the periods of vibration are much too short to be measured by conventional means, but oscillations in what is termed the radio range can be linked with a timekeeping device, producing what is known as an atomic clock. When an atom

Table 3 Composition of some atomic nuclei

Nucleus	Number of neutrons	Number of protons	Atomic number	Mass number
Neutron	1	0	0	1
Proton	0	1	1	1
Helium (in α-particle)	2	2	2	4
Carbon	6	6	6	12
Oxygen	8	8	8	16
Potassium	21	18	18	39
Potassium	22	18	18	40
Potassium	23	18	18	41
Rhenium	110	75	75	185
Rhenium	112	75	75	187
Osmium	108	76	76	184
Osmium	110	76	76	186
Osmium	111	76	76	187
Osmium	112	76	76	188
Osmium	113	76	76	189
Osmium	114	76	76	190
Osmium	116	76	76	192
Uranium	142	92	92	234
Uranium	143	92	92	235
Uranium	146	92	92	238

changes from one energy-state to another, it emits or absorbs radiation at radio wavelengths. An emitter is tuned to the frequency concerned, and the atoms are so constant in their behaviour that the clock is extremely accurate. Using a stabilised radio frequency, the clock itself, which uses a mechanical system, can be controlled and read.

One element particularly useful in this respect is cæsium. The cæsium clock, using 3·18-centimetre radio waves, is so precise that the old definition of the 'second' has been replaced by a new one, dependant upon the frequency of cæsium. In fact, mechanical time has been superseded by electromagnetic time.

Time and the atom

When an atom or an atomic nucleus is in an excited state, it will eventually fall back naturally to a lower state of excitation. With an atom this process is accompanied by the emission of light; when an atomic nucleus is involved, the process involves various phenomena which go under the general name of radioactivity.

Radioactivity

An atomic nucleus consists of two different kinds of particles, neutrons and protons, whose masses are almost equal. The *atomic number* of the element depends upon the number of protons, equal to the positive electrical charge of the nucleus. The total number of particles gives the *mass number*. Some typical examples are given in table 3.

The number of electrons moving round the nucleus is equal to the number of protons in the nucleus. The number of electrons determines the chemical properties of the atom; and since atoms are identified by means of their chemical properties, the atomic number fixes the chemical species. When a given chemical species exists with atoms of different mass numbers, the term *isotope* is introduced, and both chemical name and mass number of the

Table 4 Number of radioactive nuclei

Number	Time
N	O
N/2	T
N/4	2T
N/8	3T

atom have to be given in identification. Thus uranium-234 and uranium-238 are isotopes of the same element, uranium.

A radioactive atom, such as uranium-238, disintegrates spontaneously with the emission of a helium nucleus (alpha-particle), giving a species of a different element, thorium. It is transformation of this kind which is called radioactivity. At the end of a long and complicated series of transformations, uranium-238 becomes lead-200. Lead-200 is stable, and does not disintegrate further. Similarly, a radioactive nucleus of rhenium-187 disintegrates spontaneously and emits an electron, becoming osmium-187; this process is known as beta-radioactivity. Further details will be given later.

Not all the nuclei in a given group suffer radioactive disintegration at the same moment. In a group of N nuclei, a certain number (N/P) will undergo transformation in one second, P being the time concerned. The quantity (1/P) is known as the *constant of radioactivity*. At the end of a certain time T, the number of unaffected nuclei will have been halved; this period, T, is the *half-life*. The numbers of remaining nuclei after periods of T, 2T, etc., are given in table 4. The decrease in number follows a geometrical progression. If it is possible to determine the number N_1 at a time t_1, and the corresponding N_2 and t_2, and if the radioactive period is also known, the time interval $t_2–t_1$ follows. Radioactive periods can be measured in the laboratory; the time interval $t_2—t_1$ is then called the *time of radioactivity*.

Important problems are raised by the different methods of measuring the time intervals. Each method is based on a different set of phenomena, and each requires a different set of physical constants. It may be asked whether the student should assume the essential identity of mechanical time, energy time, electromagnetic time, and radioactive time. In the following discussion this identity will be assumed; but later in this book, under the heading of cosmology, it will be necessary to deal with theories based on the idea that mechanical time and radioactive time lead to different results with regard to the course of evolution of the universe as a whole.

Cosmical chronology

In any cosmological discussion, it is necessary to date past events as accurately as possible. In principle, a chronology can be drawn up for any evolutionary process; all that is needed is a knowledge of the laws which control it, together with information about the state of affairs at a 'starting point' from which the evolution is to be tracked. If it is possible to fix the dates of the origin of the Earth, the formation of the radioactive elements, the origin of the Galaxy and of the main groups of stars, we shall have a firm basis from which to work.

The age of the Earth

There are various methods for measuring the age of the Earth, but of these probably the best depends upon the process of radioactive decay. There are three principal families of radioactive elements: those of uranium-238, uranium-235 and thorium-232. Each ends its career as some species of lead, as follows:

Uranium-238: $\longrightarrow$ lead-206
Uranium-235: $\longrightarrow$ lead-207
Thorium-232: $\longrightarrow$ lead-208

All these three stable isotopes of lead are known to occur naturally. Therefore, however the elements were formed, there must be some natural lead which has been produced by the decay of radioactive uranium and thorium.

Of course, the original quantities of lead-206, 207 and 208 are not known, but the problem can be solved by examining rocks of different ages. These rocks were formed from a magma, and have retained the lead isotopes which have been produced throughout the Earth's history by the decay of uranium and thorium. In any given rock, the relative abundance of this lead depends partly upon the initial quantity of uranium or thorium, and partly upon the age of the rock. By using rocks of known age as reference standards, the ages of the lead-containing rocks may be deduced. Of course, the Earth itself must be older than the oldest rocks, but such differences are minor when such immense spans of time are being considered. By the radioactive-decay method, the age of the Earth works out at about 4,600 million years.

When meteorites are studied in the same way, the results are of the same order, so that presumably the Earth and the meteorites were formed at roughly the same epoch. From this, it is only one step more to derive an estimated age for the Solar System, and so providing the first and one of the most important dates in our 'cosmical chronology'. Irrespective of the method employed, it would be absurd to derive an age for the Galaxy or the universe which is less than the known age of the Earth and Solar System!

The age of the radioactive elements

It is easier, and more straightforward, to determine the age of radioactive elements. Since they virtually disappear in the course of time, they cannot have existed for an indefinite period in the past.

The Solar System, presumably formed from interstellar gas, gives information about the composition of this gas 4,600 million years ago. During the previous period, that is, from the formation of the Galaxy to the formation of the Solar System, radioactive elements had been produced, and were scattered throughout the interstellar gas. They can be accounted for only by one particular event (presumably unique), or else by a succession of events taking place in the Galaxy.

If it were possible to find out the relative abundances of two types of nuclei at the time of their original formation, then the present-day relative amounts would tell either the date of the original sole event or else the time during which the formative processes have been going on, whichever case is relevant. Of course, this information cannot be found from observation, and to decide upon the original relative abundances it is necessary to rely upon theory. To show the general nature of the investigation, it may be best to take one particular example.

Let us consider the decay of rhenium-187 into osmium-187, by the beta process (that is to say, by the emission of a beta-particle, which is nothing more nor less than an electron). The half-life of rhenium-187 is 4,500 million years. Starting from the present-day relative abundances of rhenium-187 and the two osmium isotopes, it is possible to calculate the abundance ratio of osmium-187/rhenium-187 at the time of the origin of the Solar System. The relative abundance is due entirely to the decay of rhenium-187. If we are dealing with a single unique event, its time must be put back about 6,000 million years before the formation of the Solar System: if a succession of events is preferred, all of the same kind and each producing radioactive elements, the total period involved must be something like 13,000 million years before the Solar System came into being.

In fact, it seems that the radioactive elements in the Galaxy were produced at some time between 11,000 million and 18,000 million years ago. Studies of the lead produced by uranium-238 and uranium-235 yield a similar figure. It is reasonable to suppose, then, that the radioactive elements came into existence not long after the Galaxy itself, so that by studying them we are also studying the earliest part of galactic history.

Stellar energy

Stars radiate energy, and continue to shine because they are drawing upon the energy-reserves locked up in the nuclei of their constituent atoms. Not all stars radiate by identical processes, but the general principles involved are much the same.

The Main Sequence stars, described earlîer, burn their hydrogen in a fusion reaction in which four hydrogen nuclei produce one nucleus of helium:

$$4 \text{ (hydrogen-1)} \longrightarrow 1 \text{ (helium-4)}.$$

The amount of energy released each time one gram of hydrogen is changed into helium amounts to 630,000 million joules. This enormous energy is of the same kind as that which modern scientists hope to make available by peaceful development of nuclear power; the stars are natural centres of controlled fusion, but their energy reserves are not inexhaustible. No star can transform more hydrogen than it contains.

It may be useful to go back to the analogy of the dam, described on page 87. A hydro-electric station can continue to produce power so long as water is stored up behind the dam, and similarly it may be said that a star continues to shine so long as it has enough stored energy. If the total amount of energy being sent out is known, and also the quantity of stored energy, it is possible to calculate how long the star can go on radiating.

Hydrogen makes up about 80 per cent of the Sun's mass, and this is enough to allow the present output of radiation to continue for about 10,000 million years in the future, provided that all the hydrogen is regarded as available for use. A star which is more luminous than the Sun will squander its energy reserves at a faster rate, and a very hot, blue star can hardly keep on shining with its

35 Cluster NGC 2632. In the central region of the cluster there is a 'nucleus' consisting of a quadruple star of the trapezium type, with its four components at much the same distance from each other. Ambartsumian has shown that these multiple stars are highly unstable and so must be young – a result confirmed by other studies of stellar evolution.

36 A colour-magnitude diagram for various star-clusters. The youngest clusters, such as NGC 2632, show a long Main Sequence. With an old cluster, such as M67, the upper part of the Main Sequence is missing.

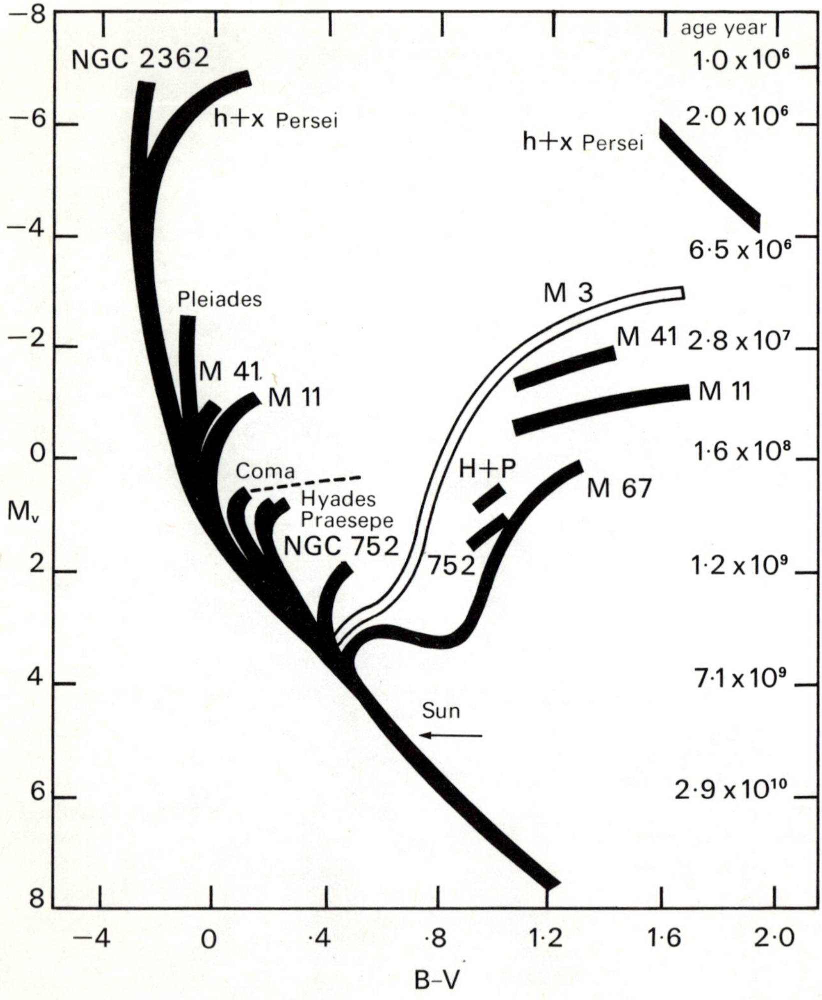

present brightness for more than about 10,000,000 years. However, these figures give no information about stellar evolution; neither do they lead to a reliable date for the origin of the stars concerned.

Stellar evolution

With stars, intrinsic luminosity increases rapidly with increasing mass. The amount of energy produced by a gram of material is greater in the cases of more massive stars, and this in turn speeds up the rate of evolution. Lone stars are not very informative, but much can be learned from star-clusters, in which the individual members are linked by gravitation into a definite system.

It is virtually certain that the stars in a cluster have a common origin, and are of about the same age, but the masses of the individual stars differ, and the most massive stars have evolved the most rapidly. In fact, a cluster offers us a sort of picture of stellar evolution. The best means of investigation is to measure the colour indexes and luminosities of individual stars. When colour index is plotted on the horizontal axis against luminosity on the vertical axis, it is clear that the distribution is not random; each cluster yields a definite pattern, and theoretical work on the colour-luminosity diagram has led to a satisfactory knowledge of the way in which a star slowly evolves along the Main Sequence.

When stars are formed, the hot gas from which they originate has been subject to violent convection currents. At this stage, a star has the same composition throughout its globe, with the same proportions of hydrogen, helium and other elements all the way from surface to centre. When the hydrogen-into-helium process begins, it can do so only where the temperature is greatest; that is to say, near the centre of the star. In most stars, there is no thorough mixing of the material, and it is the central regions which change, altering their composition and structure as the amount of hydrogen becomes less and less and the relative quantity of helium becomes greater and greater. The effects at the star's surface remain imperceptible over long periods of time; but when about 10 per cent of

37 A visual colour magnitude for the cluster NGC 188 together with field stars not associated with the cluster. The almost straight line marks the Main Sequence ; the curved line represents the stars of NGC 188. The position of the field stars, which are usually closer to us than the cluster, indicates that they are younger than the cluster stars.

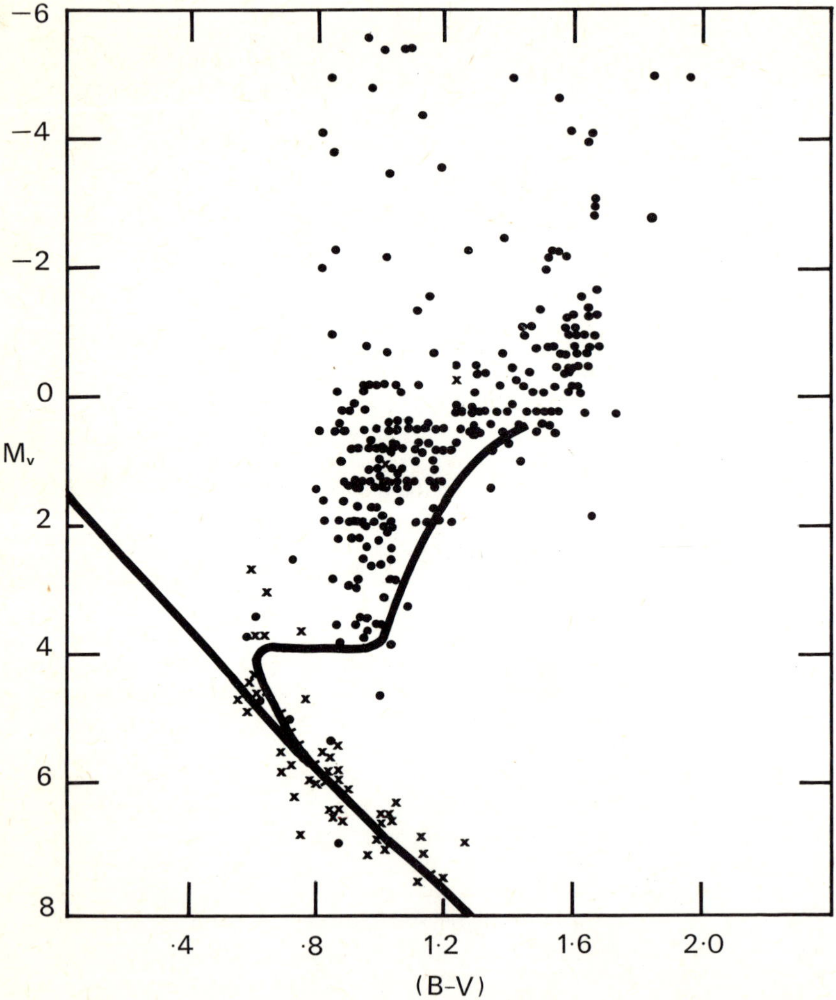

Table 5 Ages of some star clusters

Cluster	**Age in years**
Ursa Major	300 million
Hyades	800 million
NGC 2158	800 million
NGC 752	1,000 million
M 67	9,000 million
NGC 188	14,000 million

the hydrogen has been changed into helium, the star's radius and luminosity start to increase. The star becomes redder, so that its colour index increases, and the star itself moves toward the upper right of the colour/luminosity diagram.

On the diagram, it is possible to calculate theoretical curves of equal age, known as *isochrones*; it is found that these agree excellently with the curves drawn up by observation. Simply by examining the curves, the age of the cluster can then be deduced. Various clusters have been studied in this way, and some of the results are given in table 5. Of course, the clusters are not all of the same density; that known as the Ursa Major cluster is very scattered, while others, such as NGC 752, are considerably richer. As has been noted, NGC stands for Dreyer's New General Catalogue of clusters and nebulæ; M is the number of the cluster in Messier's catalogue of the eighteenth century.

The ages of the stars in the Galaxy

Empirical isochrones sweep the colour/luminosity diagram from left to right. Therefore, it seems safe to say that in general, stars which lie to the left of any given isochrone are younger than stars which are situated on the isochrone.

In this respect, the cluster NGC 188 is particularly significant. If stars in the neighbourhood of the Sun are put in on a diagram representing the stars of NGC 188, it is found that all of them lie to the left of the NGC 188 branch. It seems reasonable to suppose, then, that NGC 188 is the oldest known cluster in the Galaxy; and if the Galaxy is older still, as presumably it must be, then its age cannot be much more than that of the cluster.

The evolution of the Galaxy

The stars in our Galaxy are evolving, but so is the Galaxy itself. The next problem is to see what are the indications of this evolution, and how it affects the careers of the individual stars. The essential clues are provided by three facts, based on observation:

1 Stars evolve, and change their chemical composition, by means of nuclear reactions inside their globes.

2 Stars eject part of their material into the Galaxy – sometimes rapidly, sometimes slowly – and so modify the chemical composition of the interstellar matter.

3 Within the Galaxy, fresh stars are continually being produced from the interstellar material. This is demonstrated by the close association between the vast interstellar clouds and the youngest known star-clusters.

This latter point means that recently-born stars, produced from the modified interstellar matter, are of different chemical composition from the old stars. To see just how this process works, it is necessary to look more closely at the nuclear reactions in stellar interiors. The technical term for this is *nucleosynthesis*.

Nucleosynthesis

To give a complete description of the nuclear processes going on inside the stars would be beyond the scope of this book, but the essential stages must be noted. First, the temperature of the central regions of a star rises progressively as the star evolves. It is import-

Table 6 Nuclear reactions

Temperature (°C)	Reaction
20,000,000	4 (hydrogen–1) ⟶ helium–4
100,000,000 to 200,000,000	3 (helium–4) ⟶ carbon–12 carbon–12 + helium–4 ⟶ oxygen–16 oxygen–16 + helium–4 ⟶ neon–20
3,000,000,000	Formation of iron at equilibrium

ant to remember that reactions between atomic nuclei can become very rapid only when the temperature is high enough, and that the nuclear reactions within the star must be completed in definite phases of the star's evolution.

The first stage, already described, is the transformation of hydrogen into helium, so producing a core of helium at the star's centre. As the evolution proceeds, and the central temperature rises still more, new reactions take place in the helium core. The first of these is the transformation of helium into carbon, with the emission of very short-wavelength radiation known as gamma-radiation:

$$3\ (\text{helium-4}) \longrightarrow 1\ (\text{carbon-12}) + \text{gamma-radiation.}$$

This is followed by the transformation of carbon into oxygen and of oxygen into neon, with emission of gamma-rays:

$$\text{carbon-12} + \text{helium-4} \longrightarrow \text{oxygen-16} + \text{gamma-radiation}$$
$$\text{oxygen-16} + \text{helium-4} \longrightarrow \text{neon-20} + \text{gamma-radiation.}$$

A further increase in temperature alters the whole situation. Instead of irreversible reactions of the kind already met with, the processes become more numerous and varied; they take place so rapidly that the direct and inverse processes occur equally between the various kinds of nuclei, resulting in a state of equilibrium. Under these conditions, the nuclei which appear are those which are of a particularly stable kind: iron and its neighbours in the table of elements – titanium, vanadium, chromium, manganese, cobalt and nickel. The scale of temperatures is given in table 6.

Equilibrium conditions inside a star can be attained only when the star has reached an extreme state of contraction. For these conditions to occur in the Sun, for instance, the solar radius would have to shrink to only 1/100 of its present value; otherwise, the central temperature would not be high enough. Certainly an event of such a kind would be catastrophic. No star could maintain this temperature for very long, simply because the nuclear reactions then take place so quickly. Yet such catastrophes can be observed in nature; we call them the explosions of supernovæ.

Supernovæ

Occasionally, very powerful explosions occur in our or other galaxies; a formerly faint star suffers a tremendous outburst, and blows much of its material away into space, releasing enormous quantities of energy. It is true that few supernovæ have been seen in our Galaxy; the most celebrated are those of 1054, which produced the Crab Nebula, and two of later date, Tycho's star (1572) and Kepler's star (1604). However, many others have been observed in outer galaxies, and their general properties are reasonably well known.

At maximum, the absolute magnitude of a supernova may reach −20, so that it is then as brilliant as an entire galaxy. Studies of the way in which supernovæ vary in brightness after the initial outburst have shown that there are two definite types. Supernovæ of Type I decrease slowly in brightness after maximum; they are most often seen in elliptical galaxies and in the central parts of spirals. Type II supernovæ fall rapidly from maximum brightness, and tend to occur most commonly in the spiral arms of galaxies. It is estimated that in each galaxy, an average of three to four supernovæ is seen per thousand years, so that a supernova is a relatively rare phenomenon.

The total quantity of energy set free in a supernova explosion is enormous; it corresponds to the total energy radiated by the Sun over a period of about 10,000 million years. For a mass equal to that of the Sun, this corresponds to about half a million electron volts per nucleon. Only nuclear reactions can provide energy-sources of this order. Also, the rapidity of the explosion means that the energy is released suddenly; since reaction-speeds increase with temperature, it seems that there must be a tremendous rise in temperature just before the outburst. It is true that the contraction of a star is accompanied by an increase in the temperature near the core, but with a supernova the contraction would have to be very rapid – as rapid, indeed, as the nuclear reactions themselves. This appears to be due to a collapse of the layers of the star under their own weight, so that, broadly speaking, a supernova outburst is caused by the gravitational collapse of a star.

Formation of the heavy elements—the neutron

Whether or not the star suffers a final catastrophic outburst, and becomes a supernova, there are various other reactions which can take place at rather lower temperatures. These involve neutrons and beta-rays.

A neutron is a particle without an electric charge and with a mass slightly greater than that of a proton; taking the oxygen nucleus to have a mass of 16, with a hydrogen nucleus having a mass of 1·00812, the value for the neutron is 1·00893. It decays spontaneously, yielding a proton, an electron, and a mysterious particle called a neutrino, which has no mass and which will be described below. The process is:

$$\text{Neutron (1)} \longrightarrow \text{hydrogen-1} + \text{electron} + \text{neutrino.}$$

The neutron, whose half-life is 12 minutes, was discovered in 1931 by F. Joliot and I. Curie, who were studying the penetrative radiation emitted when beryllium is bombarded by helium nuclei (alpha particles) according to the following reaction:

Beryllium-9 + helium-4 ⟶ carbon-12 + neutron 1.

A nucleus of atomic number Z and mass number A can easily capture a neutron, to form an isotope of the same atomic number and mass number A+1. The simplest example of this, originally found in 1934, is that of a neutron being captured by a proton, forming a nucleus of deuterium ('heavy hydrogen') with the emission of gamma-radiation:

Hydrogen-1 + neutron 1 ⟶ deuterium 2 + gamma-radiation.

Beta-radioactivity and the neutrino

Some nuclei are stable, while others are unstable. If an isotope is heavy, that is to say, rich in neutrons, or light and therefore poor in neutrons, it is unstable, and changes spontaneously into a stable nucleus. If heavy, the process will be accompanied by the emission of a negative electron. If the nucleus is light, it will either emit a positive electron or capture a negative one.

This beta-radioactivity is remarkable in that the emitted electron may have any quantity of energy between zero and the maximum amount available in the transformation concerned. At first sight, this seems to contradict the law of the conservation of energy, and so in 1931 the famous physicist Pauli postulated the existence of an uncharged particle (to account for the fact that the total charge in the transformation was not affected) of very slight mass (to explain why it had not been observed). Another eminent physicist, Fermi, named this particle the neutrino. Since then, of course, its existence has been confirmed.

All fundamental particles show axial rotation, and each particle has its own 'mirror image' or anti-particle. For instance, the positive electron, or positron, is the opposite of the familiar negative electron; the positive proton is balanced by a negatively-charged anti-proton, discovered at Berkeley in 1955; and similarly, the 'mirror image' of the neutrino is the anti-neutrino. In beta-radioactivity, the emission of a negative electron is accompanied by the emission of an anti-neutrino; the capture of a negative electron goes together with the emission of a neutrino, while the emission of a positive electron involves also the emission of a neutrino.

The capture of a neutrino by matter is very difficult, and this has made it far from easy to obtain experimental evidence for the neutrino. However, final proof was forthcoming in 1955, from workers at Los Alamos. The nuclear reactions in the pile of the Savannah River atomic station release very large numbers of anti-neutrinos, and these are carried through a water reservoir, so that a few of them are captured by hydrogen nuclei:

$$\text{Proton} + \text{anti-neutrino} \longrightarrow \text{neutron} + \text{positive electron.}$$

Cadmium salts dispersed through the water allow the neutrons to be detected. The cadmium isotope which is formed becomes excited, and emits gamma radiation:

$$\text{neutron } 1 + \text{cadmium} \longrightarrow \text{excited cadmium} \longrightarrow \text{cadmium} + \text{gamma-radiation.}$$

The positive electron is captured by the negative electrons of the water, and two gamma rays are emitted. Counts were possible by means of the simultaneous detection of gamma-rays from the cadmium and the positive electron. The reactions took place at the rate of three per hour in the 500 litres of water used in the experiment.

The neutrino is so difficult to capture that to ensure its absorption by the hydrogen nuclei in the water, it would have to be sent through a column of water 250 light-years long! This property of the neutrino, the fact that its capture is practically impossible, means that it can travel for immense distances without being halted. This is of great cosmological interest.

Capture of slow neutrons

In the capture of slow neutrons, a radioactive nucleus is produced, and beta-emission follows. These slow neutrons are, therefore, always captured by stable nuclei. Starting from neon, or, more appropriately, with iron, which is relatively plentiful in an average star, nuclei of successively greater mass numbers can be built up. In the course of successive captures of neutrons, the nuclei which find the greatest difficulty in capturing neutrons become more plentiful, because they are more easily formed than disrupted. This applies to nuclei with 14, 20, 28, 82 and 126 neutrons (see table 7).

Neutrons are produced when the temperature rises because of the action of radiation upon neon. The reaction is as follows, and involves an absorption of 17,000,000 electron volts:

$$\text{Neon-20} + \text{gamma-radiation} \longrightarrow \text{neon-19} + \text{neutron.}$$

Neon-19 is transformed by beta-radioactivity, and becomes fluorine. When the temperature rises above several hundred million degrees, the numbers of high-energy photons become great enough to release a few hundred neutrons per atom of iron over a period of a few thousands of years.

Capture of fast neutrons

At slightly higher temperatures, the iron may be irradiated in a few seconds by an enormous flux of neutrons. The process is so rapid that it must necessarily occur in the region of the gravitational catastrophe which results in a supernova explosion. There is simply not enough time between two successive captures of neutrons for beta-radiation to take place, so that a nucleus will absorb neutrons until it can no longer hold them without changing its charge. Nuclei of higher and higher mass numbers are produced as the nuclei become richer and richer in neutrons. The final result of the capture of fast neutrons is the formation of the natural radioactive

Table 7 Stable elements

Number of neutrons	Atomic number	Element	Mass number
14	12	Magnesium	26
	13	Aluminium	27
	14	Silicon	28
20	18	Argon	38
	19	Potassium	39
	20	Calcium	40
28	22	Titanium	50
	23	Vanadium	51
	24	Chromium	52
	26	Iron	54
50	38	Strontium	88
	39	Ytterbium	89
	40	Zirconium	90
	42	Molybdenum	92
82	54	Xenon	136
	56	Barium	138
	57	Lanthanum	139
	58	Cerium	140
	59	Praseodymium	141
	60	Neodymium	142
	62	Samarium	144
126	82	Lead	208
	83	Bismuth	209

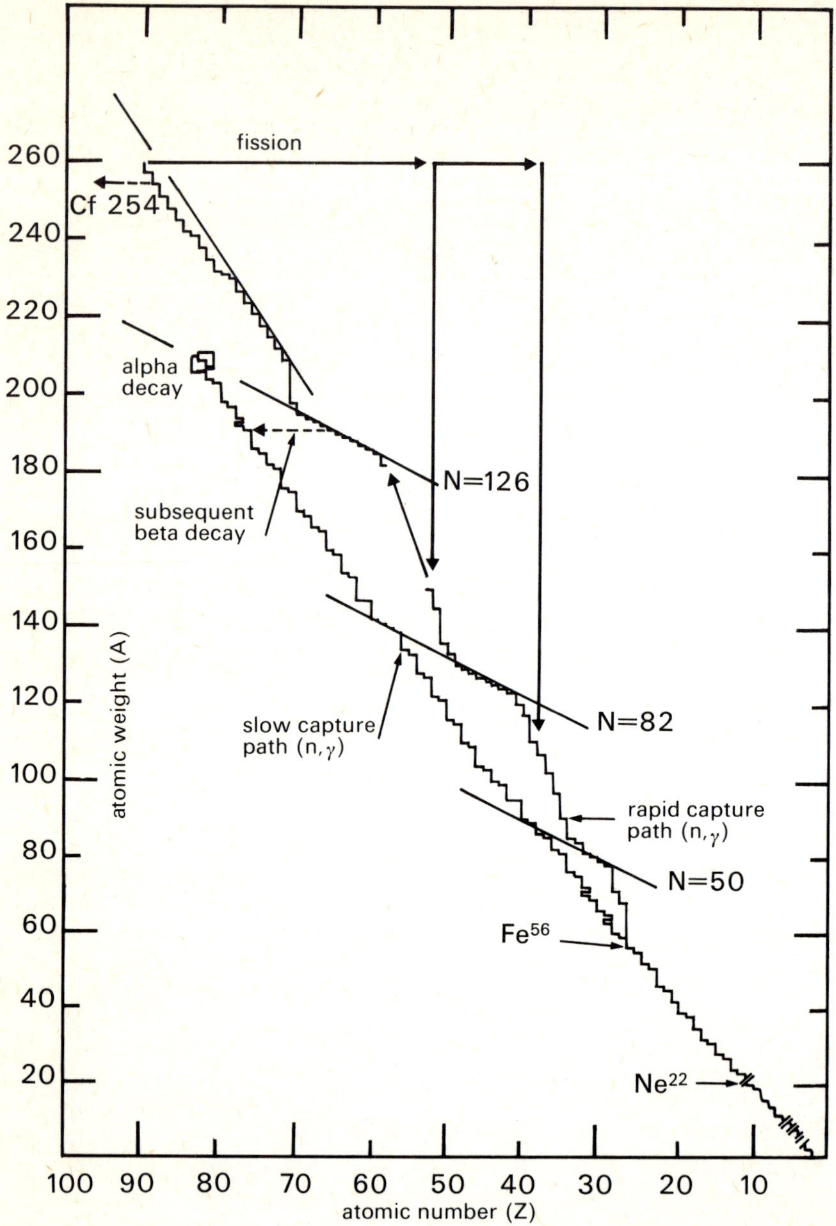
fission
Cf 254
alpha decay
subsequent beta decay
N=126
slow capture path (n,γ)
N=82
rapid capture path (n,γ)
N=50
Fe56
Ne22
atomic weight (A)
atomic number (Z)
260
240
220
200
180
160
140
120
100
80
60
40
20
100
90
80
70
60
50
40
30
20
10

38 The formation of elements by the capture of slow or fast neutrons. The diagram shows the path followed in the plan (atomic weight, atomic number) according to whether the capture is slow (so that the radioactive nuclei have enough time to decay before capturing another neutron) or fast (so that the nuclei capture neutrons one after the other until the whole nucleus becomes unstable, so that electrons and neutrinos must be emitted before the capture of neutrons can continue). The first process leads to the formation of elements such as the rare earths; the second, especially associated with supernovae, to the formation of radioactive elements.

elements. Indeed, the formation of radioactive elements cannot be explained without bringing in the capture of fast neutrons, because capture must take place so quickly that the radioactive elements are formed before they have time to decay!

It can be shown that the formation of rhenium-187 is certainly due to the capture of fast neutrons, while osmium-186 and osmium-187 are equally certainly due to the capture of slow neutrons. All this is of great importance in drawing up our cosmical chronology, as described on page 91. The initial abundance of osmium is particularly significant.

The fast processes occur in supernovæ, and an enormous flux of neutrons can then irradiate elements of the iron family. This, also, is highly significant. It is also worth noting that the formation of highly unstable elements, which soon break up by fission into several large parts, explains why some elements appear to be rather surprisingly plentiful in comparison with other members of the same element-family.

The various nuclear processes described above may be conveniently summarised in table 8, though it must be remembered that not all these processes necessarily take place in the same stars.

The chemical composition of matter in the Galaxy

Earlier in this book, something was said about the way in which spectral analysis leads to the knowledge of what elements are

39 Details to show how elements are formed. The thick line marks the slow-capture process; the dotted arrows indicate the beta-process and associated phenomena. Osmium-186 and osmium-187 cannot be produced by the fast-capture process because the series stops at tungsten-186 and rhenium-187, which have a very long half-life. Osmium-186 and osmium-187, therefore, must be due to the slow capture process.

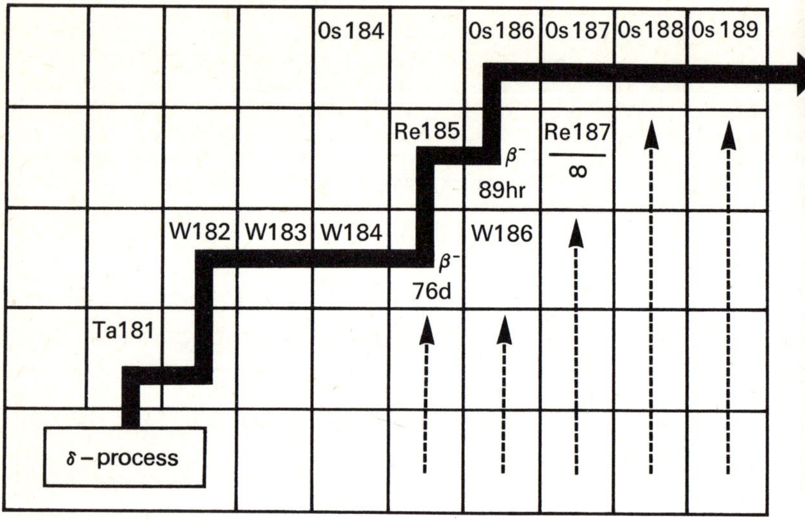

present in the outer layers of a star. Iron may be taken as a good example. It is obvious that the dark absorption lines due to iron in the star's spectrum will be strong or weak in proportion to the quantity of iron in the outer layers of the star.

To work out a value for the colour index as defined by the ratio of yellow light to blue light, or of blue to ultra-violet, it is necessary to measure the amount of light sent out by the star over a considerable range of wavelengths. In measurements of this sort, no distinction is made between the continuous background and the spectral lines. However, it is clear that colour index values derived in this way depend upon the intensities of the absorption lines present in the spectrum. An exact relationship has been established between the quantities of metals (principally iron, titanium and vanadium) present in the outer layers of a star, and the value of the colour index. Colour indexes are much easier to measure than

Table 8 Reaction temperatures and periods of reaction for the principal nuclear processes

Process	Temperature (°C)	Duration	Stars
Fusion of hydrogen	20,000,000	10,000 million years	Main Sequence
Fusion of helium	100,000,000	10,000,000 years	Giants
Capture of slow neutrons	300,000,000	10,000 years	Giants
Equilibrium process	3,000,000,000	10 months	Supernovæ
Capture of fast neutrons	10,000,000,000	10 seconds	Supernovæ

details of spectra, and have led to much important information about the chemical compositions of the stars. All this is highly significant, too, in studies of the globular clusters and the old stars in our Galaxy.

Globular clusters

Globular clusters are very beautiful objects, symmetrical in shape, and each containing tens of thousands of stars. All lie at a great distance from us. Altogether, 118 globular clusters are known, and colour index measures have been obtained for a dozen of them.

The colour index diagrams for globular clusters always show one branch, characteristic of giant stars, extending from the Main Sequence. The ages of the globular clusters can then be estimated, by the same technique as is used for galactic clusters. The reliability of the results depends partly upon the basic theory and partly upon the interpretation of the observational data, and there is bound to be a good deal of uncertainty, but it is at least safe to say that all the globular clusters are very old. The ages of Messier 13, Messier 5 and Messier 3 have been given as 22, 24 and 26 thousand million years respectively, though doubts about the theoretical model mean that these ages may have been overestimated by a factor of 2.

Colour index measures for stars in globular clusters indicate that there is a marked paucity of metals. Apparently the stars in M5 and M13 contain twenty times less iron than the Sun, while the stars in M2 and M92 contain two hundred times less iron than is found in the Sun. It follows that globular clusters must have been formed from interstellar material which was relatively metal-poor.

Old stars

On page 50 it was noted that a star which has a high velocity in the direction perpendicular to the plane of the Galaxy can go well above or below the galactic plane, whereas slow-moving stars cannot travel very far from the plane. When the velocities perpendicular to the galactic plane are correlated with colour indexes and metal abundances, a remarkable fact emerges. Stars which are richest in metals are also slow stars, moving at less than 50 kilometres per second, while metal-poor stars have velocities which may reach as much as 300 kilometres per second.

In consideration of the relationship between velocity and maximum distance from the galactic plane, it seems that metal-poor stars must have been formed at any possible distance from the plane of the Galaxy out as far as 20,000 light-years, while metal-rich stars were formed much closer to the plane of the Galaxy, over a region not exceeding 1,000 light-years in thickness.

The progressive enrichment of the Galaxy in metals, mainly because of the nuclear reactions inside the stars and the ejection of stellar material into space, makes it seem that metal-poor stars are old, whereas stars which are rich in metals are comparatively

40 The globular cluster Omega Centauri. The cluster appears circular although it is to all intents and purposes spherical.

41 Ultra-violet excess and velocity in space. The velocities of the stars in a direction perpendicular to the main plane of the Galaxy are associated with metal abundance in their atmospheres. The ultra-violet excess is a measure of this.

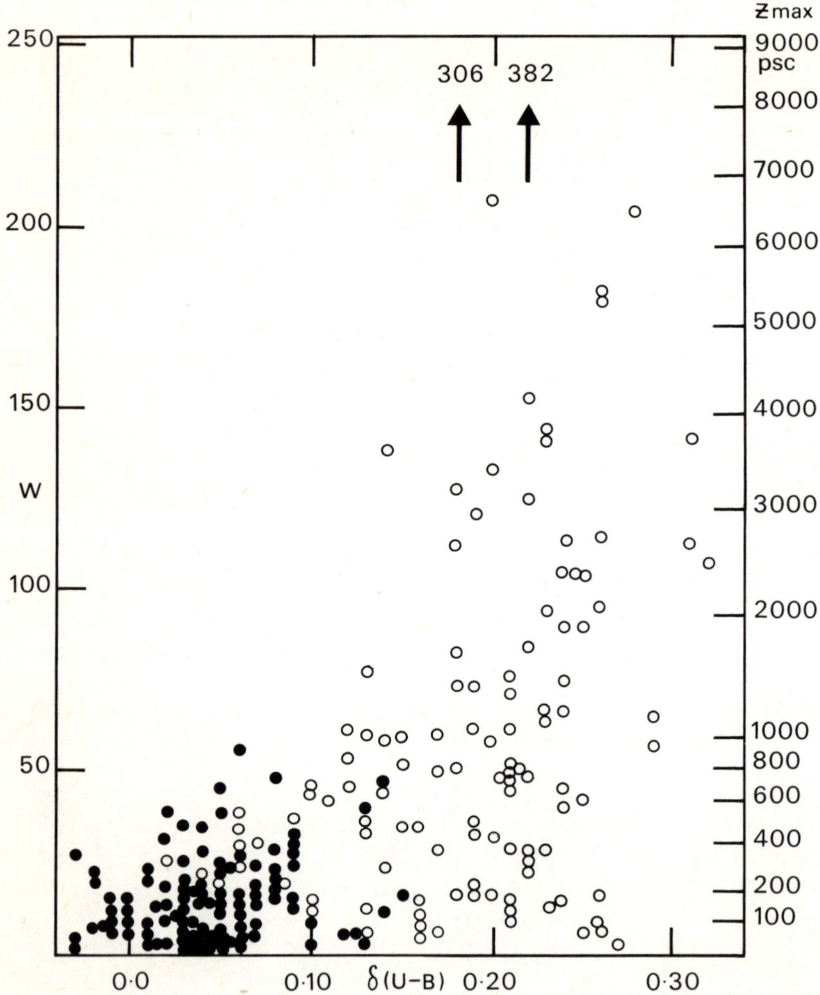

young. Old stars were formed at a period when the main interstellar material was spread out to great distances on either side of the galactic plane, while the younger stars were produced when, as at present, most of the material was confined to a region relatively close to the plane. During the time when old stars were being formed, the Galaxy must have contracted toward its equatorial plane.

Stars with large positive colour indexes move in very eccentric orbits. Paths of this sort can occur only when the star's velocity in the direction of the galactic centre is comparable with its velocity relative to the perpendicular to the plane of the Galaxy. The material from which old stars were formed must thus have had a considerable velocity in the direction of the galactic centre, and contraction toward the plane should be accompanied by contraction toward the centre. If the rate of contraction were roughly the same as the orbital velocity, it can be shown that the contraction would have taken from 100 to 200 million years. Globular clusters must have been formed very quickly at the start of this phase of evolution of the Galaxy, and were left behind, at a great distance from the galactic centre, as the gas contracted.

The sizes of the eccentric orbits of old stars show that the original cloud must have been about 300,000 light-years in thickness, since when it has contracted to one-tenth of this size in the equatorial plane and one-twenty-fifth in the perpendicular direction. The great contraction toward the galactic plane can be explained only by the effects of the general rotation of the whole system.

Conclusion

The ideas of time and chronology are fundamental in any understanding of the universe. At the start of the present chapter, it was pointed out that astronomical events develop over periods of time, and that each kind of phenomenon has its own particular time-scale. Motion, energy transformation, radiation and radioactivity are entirely independent of each other, and each one may be used to define what we call 'time'. Moreover, it is a fundamental postulate that 'time' as defined with reference to any one of these phenomena is always the same time, so that the chronology of a phenomenon C is the same whether it has been worked out according to the time-scale based on phenomena A or B. It is true that some physicists have cast doubts upon the coincidence of the different time-scales, and this will be discussed in more detail later, but it must be said that at the moment there is no valid method of deciding who is right and who is wrong.

For the moment, at least, it will be best for us to keep to the common time-scale in summarising the points that have been made here. There can be no doubt that evolution is in progress all around us. The Galaxy has not always been in the state that it is in today; the stars in the Galaxy are born, change, and so far as most are concerned, probably end by disappearing from our view to wander, dark and invisible, through the Galaxy.

Studies of star-clusters lead on to estimates for the ages of some systems. The oldest of these systems have an age of perhaps 15,000 million years, while the youngest date back for less than one million years. Observation even indicates that star-formation is taking place before our eyes; the most striking example is that of the star FU Orionis, which became luminous in 1937 over a period of less than three months in a position where there had previously been no definite star and where there had been nothing more than some very dim object. The researches now going on will presumably lead on to a better knowledge of the very earliest stages of a star's evolution.

Various remarkable events can happen during stellar evolution. In some cases, marked chemical transformations due to nuclear reactions modify the composition of the stars concerned, and a large fraction of the material is hurled out into interstellar space, thereby causing a slow but steady change in the chemical composition of the interstellar material; metals become more and more abundant. It is likely that activity of this kind was more intense in the early history of the Galaxy than it is now. A few stars seem to suffer violent outbursts which end as supernova explosions; it is probable that during this process, radioactive elements are regenerated. Of course, nothing certain is known about the time when the regeneration process and the formation of the radioactive elements began, but it may have been something like 18,000 million years ago.

The rapid evolution of the Galaxy soon after its formation affected both the chemical composition and the form. Flattening soon took place, so that old, metal-poor stars are situated well away from the galactic plane, while young, metal-rich stars are found much closer to the plane. And without being too hard and fast about it, there is reason to think that the distinction between Populations I and II are associated with age and composition. Stars of Population I are young and metal-rich, while stars of Population II are old and metal-poor.

At the present time, all these problems are being closely studied by astronomers; more is being learned about stellar structure and evolution, and about the chemical composition of stars and nebulæ. Just as the study of the Sun is of fundamental importance in finding out more about other stars, so the study of our Galaxy is vital in any attempt to understand other galaxies. Yet the almost incredible varieties of forms presented by the outer systems lends support to the view that some of the mysteries of our Galaxy will be solved when we have been able to interpret the other galaxies lying far away across space.

4 Space and the galaxies

Photography shows that there are enormous numbers of galaxies, diffuse objects of regular outline. The brightest of them are also the closest to us, and can be resolved into stars. The more remote galaxies cannot be so resolved, even with our most powerful instruments, so that they look like dim, nebulous patches.

Even a cursory study of the photographs shows that the galaxies may be divided into two main classes. Some galaxies, rich in blue stars and interstellar gas, are of spiral structure; they are flattened and give clear indications of quick rotation. Others, poor in blue stars, have no spiral structure and are much less flattened, but they give the impression of an ellipsoid in a state of rotation, and are called elliptical galaxies. Actually, the classification of galaxies is much more complex than might be thought at first sight, and this is one of the problems to be examined in the present chapter. First, it will be helpful to list a few of the questions to be tackled.

The positions in space of the galaxies must be determined. This, of course, brings us back to the old problem of distance-measuring, but geometrical and kinetic methods, so useful in estimating the distances of stars in our own Galaxy, are of no use for the external systems; the distances involved are much too great, and the proper motions of galaxies are too slight to be measured at all. Less direct methods have to be found, based chiefly upon the intrinsic luminosities of the galaxies concerned.

The next problem is that of classification of the different types of galaxies, bearing in mind their stars, their gas, and their shapes. With stars, our physical knowledge is good enough to provide a sound basis for classification, but this is not true for the galaxies. So little is known at present that there is no choice but to follow methods of analysis resembling the methods once used by naturalists in classifying plants and animals.

A third problem, serving as a kind of introduction to an overall study of the universe, involves finding out the distribution of the galaxies in space. This does not mean the general distribution all through the universe, but refers to the grouping of galaxies in systems. Moreover, the masses of the galaxies must be investigated,

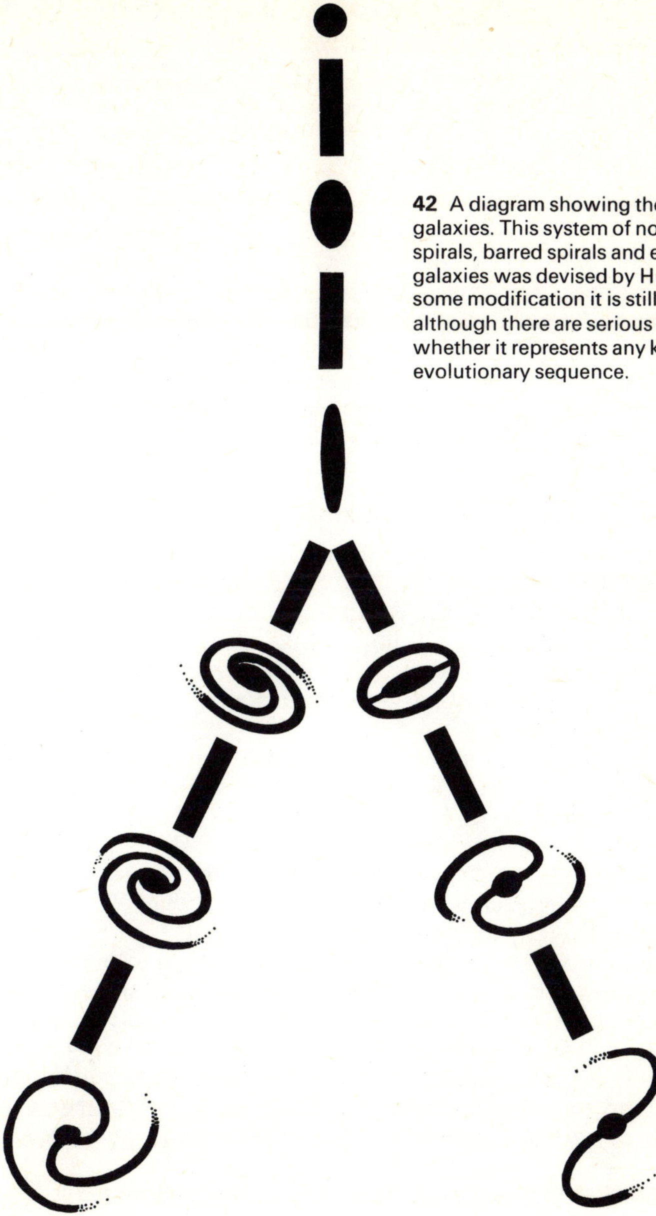

42 A diagram showing the types of galaxies. This system of normal spirals, barred spirals and elliptical galaxies was devised by Hubble. With some modification it is still used, although there are serious doubts whether it represents any kind of evolutionary sequence.

43 The 120-in reflector at the Lick Observatory, USA. This has a skeleton tube and a conventional mounting. It is one of the most recent of the great reflectors and is second in size only to the 200-in reflector at Palomar.

since this leads on to a knowledge of the mean density of matter in space.

The development of the special science known as cosmology, to be discussed in the last section of the present book, depends largely upon a sound knowledge of the galaxies. Summing up, it may be said that we need information about: (a) the distances of the galaxies, (b) luminosities, (c) classification of different types of galaxies with respect to form and content, (d) distribution of the galaxies in space, and (e) the masses of the galaxies.

The distances of the neighbouring galaxies

There are several ways in which the distances of relatively nearby galaxies can be measured. In each case, certain bodies in the galaxy concerned are studied, and their intrinsic luminosity measured. As soon as the real luminosity is known, together with the apparent brightness, the distance of the object can be worked out, and this, of course, also gives the distance of the galaxy in which the object lies.

Effectively, the relationship between apparent magnitude and absolute magnitude is a clue to the distance of the object. When the magnitude differences are, respectively: 0, 1, 2, 3, 4, 5, 6, etc, the distances involved are, in parsecs, 10, 16, 25, 40, 63, 100, 160, and so on. (One parsec is equal to 3·26 light-years.) For example, a difference of 35 magnitudes corresponds to a distance of 100,000,000 parsecs, or about 326,000,000 light-years.

Something has already been said about the period-luminosity law of Cepheid and RR Lyræ variables, according to which the distances of these variables can be found as soon as their periods of fluctuation are known. Clearly, the method can be applied to the galaxies. When a suitable variable has been discovered inside a galaxy, and its period measured, its distance can be found; and this leads on to the distance of the galaxy which contains the variable.

Novæ are also very useful. They are important in many ways, and

represent stellar outbursts on a grand scale, though they are much less violent than supernovæ, and after its outbreak a normal nova returns to something like its former condition. Many novæ have been observed in our Galaxy, and they occur in other systems also. Though the absolute magnitude of any individual nova may not be known, statistical methods show that in our Galaxy, the mean absolute magnitude of a nova at maximum is −7·5. If a nova is seen in an external system – the Andromeda Galaxy, for instance, where many novæ have been discovered – it is reasonable to assume that the maximum absolute magnitude is also −7·5. Comparison of this value with the observed apparent magnitude then gives the distance.

Over a hundred globular clusters are known to be associated with our Galaxy; each is symmetrical, and each contains a large number of stars. Their distances have been measured (by means of their RR Lyræ stars), and their absolute magnitudes have been calculated; the value adopted is −7·5. Globular clusters are also known to be associated with other galaxies; and if these too have absolute magnitudes of about −7·5, their distances can be worked out in the same way as for novæ.

Another method can be used for galaxies which are close enough to be resolved into stars. The brightest of the red stars in galactic globular clusters are compared with the brightest of the red stars in the galaxies, and it is assumed, very reasonably, that the absolute magnitudes are the same. Comparison of the absolute magnitudes with the apparent magnitudes then leads to values for the distances involved.

These various methods are applicable only to the relatively few galaxies close enough to be resolved into stars. The results are as follows:

1 *Small Magellanic Cloud.* Both the Large and Small Magellanic Clouds, which are situated in the southern part of the sky some distance from the Milky Way, are satellite systems of our Galaxy. It is required to find out the *distance modulus*, which may be

44 The dome of the 102-in reflector at the Crimean Astrophysical Observatory in the USSR. This photograph was taken in 1960, when the dome was nearly complete. It is the largest telescope in the Soviet Union and is surpassed in size only by the Palomar 200-in and the Lick 120-in reflectors.

defined as m – M, in which m is the apparent magnitude and M the absolute magnitude. For the Small Cloud, the distance modulus, as obtained by various methods of investigation, is as follows:

METHOD	DISTANCE MODULUS, m – M
Short-period variables	19·1
Novæ	18·7
Novæ 15 days after maximum	18·8
Red giants in globular clusters	19·1

The mean value is 18·9. When the effects of absorption are allowed for, adding about 0·5 to the apparent magnitudes of stars in the Small Cloud, the distance is found to be 55,000 parsecs.

2 *Large Magellanic Cloud.* The distance modulus for the Large Cloud is measured in the same way, with the following results:

METHOD	DISTANCE MODULUS, m − M
Short-period variables	19·2
Red giants in globular clusters	19
Classical Cepheids	18·6
Novæ	19
Novæ 15 days after maximum	18·4

There is some disagreement here, and the correct result may be anything between 18·7 and 19, corresponding to an error of roughly 30 per cent in the distance estimate. If the absorption is held to add 0·4 to the apparent magnitude, the distance of the Large Cloud is found to be 45,000 parsecs, so that it is appreciably closer than its smaller neighbour.

3 *The Andromeda Spiral, Messier 31.* Many methods have been used to measure the distance of this beautiful northern galaxy, whose apparent magnitude is 4·33. The results as obtained in different ways are:

METHOD	DISTANCE MODULUS, m − M
Photographic measures of the overall magnitudes of globular clusters	24·8
Photoelectric measures of the overall magnitudes of globular clusters	23·9
Brightest stars of Population II (red giants)	24·3
Classical Cepheids	23·9
Novæ	24·6

The mean value is 24·4. Absorption increases the apparent magnitude by about 0·6, and so the distance was found to be about 550,000 parsecs or 1,800,000 light-years, with an absolute magnitude of about −20. Very recent studies have increased this slightly, and the latest value for the distance of the Andromeda Spiral is 2,200,000 light-years.

45 An unstable system of multiple galaxies; Stephan's Quintet.

Table 9 The most luminous stars in various galaxies

Star	Galaxy	Absolute magnitude	Distance modulus
N 12 VI Cygni	Our Galaxy	−9·8	—
Star in Scorpio	Our Galaxy	−9·4	—
Star in Orion	Our Galaxy	−8·8	—
HDE 26970	Large Magellanic Cloud	−9·8	19·2
HDE 269781	Large Magellanic Cloud	−9·5	19·2
HD 33579	Large Magellanic Cloud	−10·1	19·2
HD 7583	Small Magellanic Cloud	−8·8	19·2
HD 6884	Small Magellanic Cloud	−8·5	19·2
Mean value	Andromeda Spiral, M31	−8·9	24·6
Mean value	Triangulum Spiral, M33	−8·6	24·5

Highly luminous stars

To take matters a stage further, new distance-markers are required, and they are found in the highly-luminous blue stars contained in almost all galaxies (excluding certain dwarf systems consisting entirely of Population II). The absolute magnitudes of these highly-luminous stars may be worked out, whenever the distances of the galaxies in which they lie have been measured in other ways. The results are given in table 9.

If similar stars of high luminosity are found in remote galaxies, they can be used as standards of reference for distance-finding. They can be seen out to greater distances than the short-period variables, because they are much more luminous. Taking the absolute magnitudes for the brightest blue stars in galaxies as −8·5, the following distance moduli are obtained:

GALAXY	DISTANCE MODULUS, $m - M$
M81	27
M87	30·8
M100	29·3
NGC 4321	30·8

One difficulty about this method is that the exceptionally luminous stars may be surrounded by bright gaseous nebulosity, not easy to recognise. If the nebulosity is unwittingly lumped together with the star, the resulting magnitude error can be as much as 1·8. The only remedy is to compare photographs taken in blue and in red light, so that the blue supergiants can be distinguished from any bright nebulosity nearby, and the required corrections can be made. The values of the distance moduli for M87 and NGC 4321 involve adjustments of this kind.

Clusters of galaxies

Sky photography shows that galaxies often group together in systems which are called clusters of galaxies. Various clusters have been found, of which those in Virgo, Ursa Major, Coma and Boötes are particularly notable. Some are comparatively close; others, such as the cluster in Hydra, are extremely remote by any standards.

The number of galaxies detectable in a cluster depends upon the limiting magnitude of the instrument used. For instance, the rich cluster of galaxies in the constellation of Coma is shown to have 654 members on a plate taken with the 18-inch Schmidt telescope at Palomar, but as many as 10,724 members on a plate taken with the 48-inch Schmidt. The first instrument showed galaxies in the magnitude range from 13·2 to 16·5, but the 48-inch could reach down as far as magnitude 19·0.

The clusters include galaxies of all types: elliptical, spiral, and in great numbers, flattened systems without spiral structure, and devoid of interstellar material. These latter systems are classed as galaxies of type S0. In addition, the clusters contain objects of different brightness, so that attempts can be made to draw up a statistical analysis of the magnitudes of individual galaxies in a cluster. The number of galaxies with magnitudes between m and m + 1, m + 1 and m+ 2, m +2 and m + 3, and so on, is counted, and what is termed a *luminosity function* is obtained. The similarity of the luminosity function for different clusters is quite remarkable. Another method is to arrange the separate galaxies in order of decreasing brilliancy, and then consider, for instance, the brightest ten, comparing them with the ten brightest members of a separate cluster. Here, too, there seems to be surprising uniformity between the clusters. In twelve cases, for instance, the difference between the magnitude of the brightest separate galaxy and the tenth in order lies between 0·74 and 1·73, with a mean value of 1·29.

All these results seem to show that the clusters of galaxies are of the same basic kind, and have the same basic origin. In the following discussions, it will be assumed that the mean absolute magnitude for the ten brightest galaxies in a cluster is always the same. This may not be strictly true, but it is not likely to be very wide of the mark. Of course, as soon as the absolute magnitudes of the brightest galaxies in a cluster have been found, the distance of the cluster itself can be worked out. So far, this process has been carried out for 18 clusters.

The absolute magnitude of the Andromeda Spiral is −20·3, while for NGC 4321 and M87 in the Virgo cluster the values are respectively −20·91 and −20·96. The mean absolute magnitude for the brightest galaxy in a cluster can be taken to be −21. With this value, the distance modulus for the Hydra cluster is 40. Neglecting any corrections, of which more will be said later, this corresponds to an approximate distance of 1,000 million parsecs, or 3,000 million light-years.

The red shift

The first observations of the spectra of galaxies were made by Slipher in 1912. On his photographic plates, Slipher was able to

46 NGC 205 photographed with the 200-in Hale reflector at Palomar. This is an elliptical galaxy and is a companion or 'satellite galaxy' to the Andromeda Spiral, M31. It is probably true to say that NGC 205 and the second satellite galaxy bear the same relationship to M31 as with the two Magellanic Clouds in the case of our own Galaxy, though the types are not the same.

47 The Cluster in Corona Borealis.

measure the radial velocities of the galaxies, by means of the Doppler effect, and by 1925 he had obtained results for 49 galaxies. Most of the Doppler shifts were toward the red end of the spectrum, indicating velocities of recession. Using these results, E. E. Hubble, in 1929, was able to establish a definite relationship between distance and the velocity of recession. Two years later, Hubble and his colleague M. Humason obtained important results from studies of the clusters of galaxies, and drew up the so-called Hubble-Humason law linking distance with radial velocity. The value which they obtained was 550 kilometres per second per million parsecs.

A good many corrections have had to be made to this value, mainly because of errors in the absolute magnitudes of the objects used in the compilation of the Law. The constant H of Hubble and Humason's relationship is fixed nowadays in three stages:

1 The mean apparent magnitude of the ten brightest galaxies in a cluster is determined, and the amount of the red shift in the spectral lines is measured. Using the value for the red shift as measured, it is possible to determine the section of the spectrum which is seen and to calculate the apparent magnitude that the galaxy would have in the absence of any red shift. To some extent, of course, the correction depends upon the spectral characteristics of the galaxy, but this can be allowed for. For a red shift corresponding to a recession of 60,000 kilometres per second, the correction has a value of 0·94.

2 Taking the red shift, expressed as a fraction of the velocity of light, as a function of the mean apparent magnitude which has already been found (corrected for the effects of the red shift and also for the absorption experienced by the light when crossing our Galaxy), a relation can be found between the corrected apparent magnitude and the red shift. This leads to an exact relationship between red shift and distance.

3 The scale is then fixed by means of the absolute magnitude of the ten brightest galaxies in a cluster.

Take, for example, the cluster in Hydra. The radial velocity is 0·2 of the velocity of light. The next steps are to apply corrections of 0·94 because of the red shift, and 0·49 on account of absorption in our Galaxy. With an absolute magnitude of −21, the derived distance modulus is 38·77, corresponding to a distance of 520,000,000 parsecs or 1,700,000,000 light-years.

What emerges from these investigations is a linear relationship according to which the radial velocity of recession increases by 100 kilometres per second per million parsecs. On this reckoning, the distance of the Hydra cluster would work out at 600,000,000 parsecs or 1,950,000,000 light-years. The value for the Hubble-Humason Law adopted at the moment (100 kilometres per second per million parsecs) is naturally not definitive, and will no doubt be

48 The distribution of clusters of galaxies over a certain region of sky (after Zwicky). Each figure corresponds to a radial velocity interval: (a) below 15,000 km/sec; (b) 15,000 to 30,000 km/sec; (c) 30,000 to 45,000 km/sec; (d) 45,000 to 60,000 km/sec; (e) over 60,000 km/sec.

improved in the future, but it serves very well in estimating the distances of very remote galaxies, and it has to be used when we are trying to reach out to the furthest limits accessible to us.

Classification of galaxies

The preliminary classification of galaxies given at the start of this chapter is, of course, inadequate. Several methods of more detailed classification are possible. The first of these, used extensively by Hubble, depends upon the morphological aspects of the galaxies as observed optically. Another is based upon spectroscopic studies, particularly in the characteristics of the radio emission.

Optical appearance

As has been noted, galaxies may be classified as either elliptical or spiral, but more properly there are four main classes to be considered; elliptical (type E), lenticular (S0), spiral (S) and irregular (I). This is Hubble's classification.

The so-called elliptical galaxies range from globular forms to very elongated lenticular shapes. In general, they show no structural details other than the small, bright nucleus, around which the glow of the galaxy fades away in each direction until it is lost against the dim luminosity of the night sky,

Lenticular galaxies are characterised by a bright nucleus lying at the centre of the lenticular disc, which has a fairly sharp boundary surrounded by a feeble, diffuse envelope. In their generally flattened forms they bear some resemblance to the spirals, but they contain very little interstellar material, and there is of course no spiral structure.

In the spiral galaxies, there are various spiral arms which may be open or tightly-wound. Hubble distinguished two families: the normal spirals, with arms coming out tangentially from the central bright nucleus, and the barred spirals, in which the arms come from the ends of a bar crossing the bright nucleus. The amount of

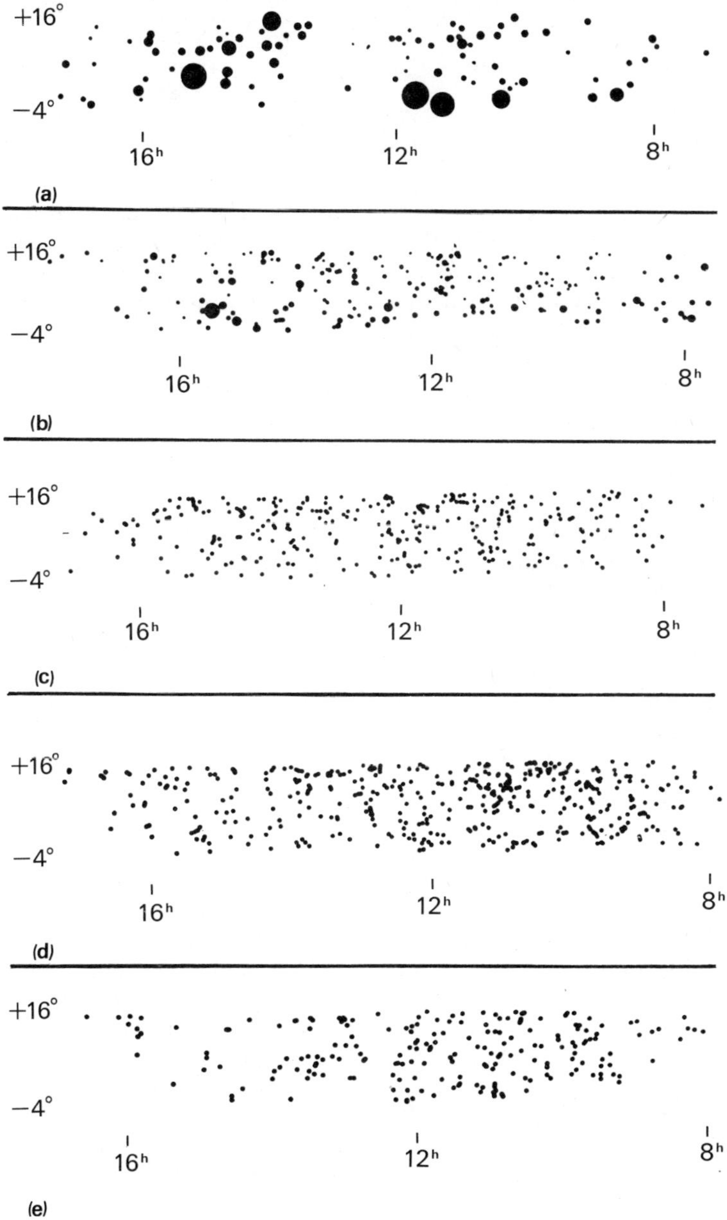

+16°
−4°
16h
12h
8h
(a)
+16°
−4°
16h
12h
8h
(b)
+16°
−4°
16h
12h
8h
(c)
+16°
−4°
16h
12h
8h
(d)
+16°
−4°
16h
12h
8h
(e)

opening of the arms is shown by the letters a, b or c; thus Sa spirals are tightly-wound, while with Sc systems the spiral is very loose.

Irregular galaxies, as their name implies, are structureless. Into this class Hubble put all the galaxies which did not seem to fit into any of the other types.

Strictly speaking, the classification of galaxies cannot be limited by a simple division into four large classes. There are intermediate types, and it is even possible that there is a continuous gradation from the ellipticals right through to the irregulars. However, a classification of such a kind would be intelligible only if we could solve the problems of the forms of the galaxies, and it must be admitted that no such theory is available as yet, although some basic ideas have been proposed.

Spectral classification of galaxies

No classification can be based upon spectral results only; it must also be morphological. The following are the main types:

1 *Spirals*. These are classified by means of a comparison between the brightness of the central region, and the total brightness of the galaxy. The centre of a galaxy will show a spectrum which is a combination of the spectra of all the various stars in the region, and it is bound to be somewhat confused, but it does give information about the main components of the area under inspection.

At this point it is worth recalling that before colour index measures became so widely used, the stars were always classified according to the dark absorption lines in their spectra (a system which, of course, is still in use). The main types were:

B: bluish-white stars, with prominent helium lines in the spectra.
A: white stars, with prominent hydrogen lines.
F: yellow stars, with prominent spark-lines of metals.
G: yellow stars, with arc-lines of metals.

K and M: orange-red stars, with spectral lines due to the molecules of metallic oxides.

By analogy, the composite spectra of the centres of galaxies are indicated by the letters a, f, g and k. The use of these letters is enough to distinguish between the spirals, because the relative brightness of the centre to the overall brilliancy is closely related to the spectral type. There is, therefore, an approximate relationship between Hubble's types Sa, Sb and Sc, classified by the increasing opening of the spiral arms, and the spectral types of the centres, as follows:

Sa	k S
Sb	g S – gk S
Sc	a S – fg S

with types gk and fg being intermediate between g and k, and f and g, respectively.

2 *Ellipticals.* The classification is essentially the same as Hubble's.

3 *D-galaxies.* These galaxies are distinguished by an elliptical nucleus surrounded by an extended envelope. Some of them would have been classed as S0 in Hubble's system so far as their optical aspect is concerned, but the correspondence between the two classifications ends there, because all the S0 galaxies are very flattened, while no flattened D-type galaxy has ever been observed. There is a large range of luminosity, with the brightest galaxies being at least ten times more luminous than the feeblest. What may be called supergiant D-galaxies, found near the centres of some clusters, have diameters three or four times greater than those of lenticular galaxies of the same cluster.

4 '*Dumb-bell*' *galaxies.* These are analogous to the D-galaxies, but each has two nuclei in the same envelope. This may be an extreme

case of multiple, tightly-packed galaxies in which there are only two equal components.

5 *N-galaxies*. A galaxy of this class has a bright uncleus, stellar in appearance, which is responsible for nearly all the light coming from the system. It is surrounded by a faint nebulous envelope, which does not stretch outward for very far.

6 *Quasi-stellar sources* (Qs). These remarkable objects, also known as quasars, will be described in detail later. Whether they can be classed with other galaxies remains to be seen. They are stellar in aspect, but sometimes accompanied by faint nebulous wisps. They emit intense ultra-violet radiation, and the emission lines in their spectra are broad.

Radio sources

In 1946 Hey, Parsons and Phillips were studying the continuous radio emission from the Milky Way when they discovered some small discrete sources, which did not agree in position with bright stars. More than 2,000 of these discrete radio sources have now been catalogued. To identify them with optical objects is as difficult as it is important; the optical objects are usually faint and, to make matters worse, the position of a radio source can seldom be fixed as accurately as with an object which can be observed visually – simply because a radio telescope is very much inferior in resolving power. At the present moment, a large number of radio sources have been conclusively identified with optical objects, many of which are galaxies. No doubt the list of identifications will grow quickly in the future.

The power of a radio source may be measured either by considering the total amount of energy sent out in the radio range, or else by means of a 'radio magnitude' based on the intensity at a wavelength of 190 centimetres. If both optical magnitude and radio magnitude are compared, it becomes clear that galaxies may be divided into two classes: weak radio sources and intense radio sources. This division is confirmed by examination of the total radiative power. If the strength of emission in the radio range is plotted, and different symbols are used for the different types of galaxies, a sharp separation is found between strong and weak sources at about 10^{41} ergs per second. However, the separation between strong and feeble radio sources undoubtedly takes place at a power lower than this.

Radio sources can also be classified by means of a *spectral index*, which shows the manner in which the amount of emitted radiation varies as a function of the wavelength. In this way, sources in which the radiation is of thermal origin can be distinguished from those in which the emission is of the synchrotron type.

Optical appearances of radio sources

All the spiral galaxies belong to the group of weak radio sources. The strong radio sources are galaxies of type D dumb-bell, N and Qs, with radio sources of type D showing the greatest range of radio power (in the ratio of 10,000 to 1).

The most powerful and luminous of all these objects are the quasars, or quasi-stellar objects. In the radio range, their power ranges from $2{\cdot}10^{44}$ to $2{\cdot}10^{45}$ ergs/second, or from 50,000 million to 500,000 million times the power sent out by the Sun. Table 10 gives the radio power, absolute magnitude, distance and radio size for four quasi-stellar objects for which the speed of recession is known fairly satisfactorily. The main details have been calculated on the basis of Hubble's constant, taken as 100 kilometres/sec per megaparsec. The high luminosity of these objects is particularly notable; it amounts to almost 100 times the brightness of the brightest normal galaxies. The power sent out in the optical range by the quasi-stellar sources is truly enormous; 300,000 million times that of the Sun in the case of 3C-47, as much as 2,000,000 million times that of the Sun for 3C-273. To make the situation even stranger, it is found that some of the objects, including 3C-47

Table 10 Data for quasi-stellar sources

Name	v (km/sec.$^{-1}$)	L (radio)	Mv		R	d (radio)
				megaparsecs		kiloparsecs
3C-47	127,500	$1{\cdot}5 \times 10^{44}$	−23	1,280	250	250
3C-48	110,100	$4{\cdot}7 \times 10^{44}$	−25	1,100	3	3
3C-147	163,500	2×10^{45}	−25	1,640	10	10
3C-273	47,400	3×10^{44}	−26	470	40	1

and 3C-273, are variable; their fluctuations can amount to 45 per cent of the total brilliancy over a period of time of only a month. This represents a variation in power of several hundred thousand million times the luminosity of the Sun, and at the moment we have to confess that we do not know what causes the fluctuations. In addition to a continuous spectrum, a quasi-stellar source shows in the optical range an emission spectrum of great intensity, and in comparison with the normal galaxies the spectrum of a quasar is distinguished by the fact that a great excess of ultra-violet radiation is emitted.

Incidentally, it is worth noting that there are no very flattened galaxies among the strong radio sources, while weak sources, including most of the spirals, are strongly flattened. This alone suggests that there are fundamental differences between the strong and the weak radio sources.

Radio properties of different optical types

A weak source always occurs as a single source centred upon a galaxy; its diameter may be smaller than, or comparable with, that of the galaxy concerned. The double structure characteristic of strong sources is never found with weak ones. It seems, therefore, that the central regions of spiral galaxies are responsible for the radio emission, and the cause may well be expansion and violent turbulence of the gas in the central region, as with our own Galaxy.

In the radio range, the strong sources show a characteristic double structure, so that in most cases the radio waves come from two regions placed symmetrically to either side of the galaxy concerned (Lequeux). Some of the strong single radio sources probably do not betray their double nature simply because our instruments are not sensitive enough to resolve them. However, not all the strong sources show the same optical properties. The weakest D and E galaxies are single sources, and so are most of the dumb-bells. It is quite on the cards that most of the D, E and dumb-bell galaxies really are single sources, which would mean that double sources would be in an overall minority. The structure of N-galaxies in the radio range is decidedly complex, and the quasi-stellar objects (Qs) show all possible aspects – single, double, single but surrounded by a halo, and so on. It may be that these differences are due to nothing more than differences in the progress of evolution, but it is impossible to be sure.

The radio radiation of the strong sources is essentially due to the synchrotron process, caused by high-energy electrons. The problem of the energy supply of the intense radio sources is so important that it will be examined in more detail later.

Multiple galaxies

No classification would be complete without including the multiple galaxies. These are systems of two or more galaxies in association, and often connected by a bridge of faintly luminous material stretching across the space between them.

To understand why the multiple galaxies are so significant, it will be helpful to return to elementary celestial mechanics, and say something about the properties of bodies in motion. The motion of two bodies can be either elliptical, hyperbolic or parabolic. Bodies which are moving in ellipses relative to each other will be permanent neighbours in space, while with parabolic or hyperbolic movement the association will be no more than temporary; once they begin to separate, the two bodies will continue to move apart

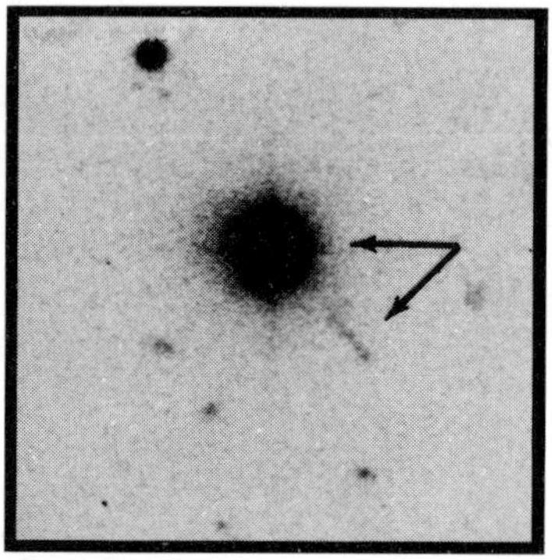

3C 273 Qs

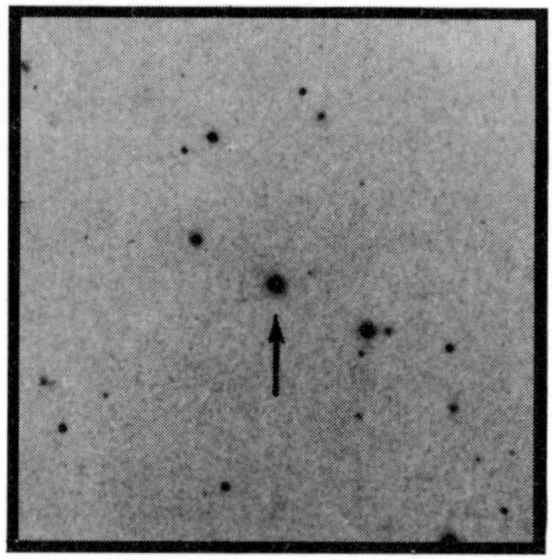

3C 445 N

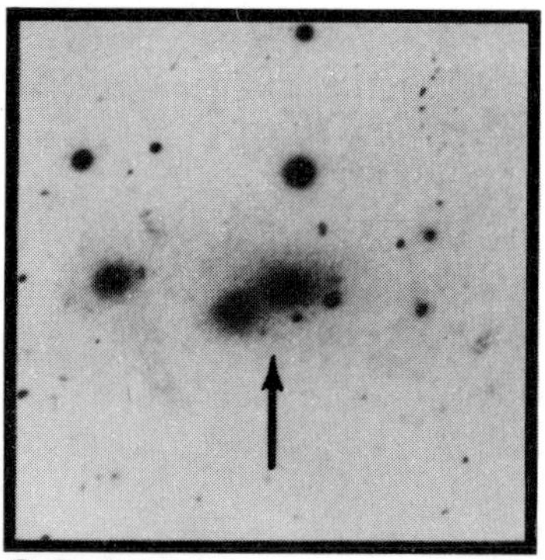

3C 310 db

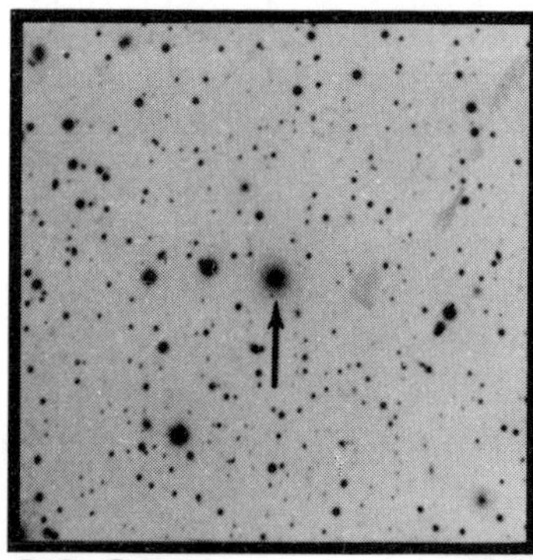

3C 353 D2

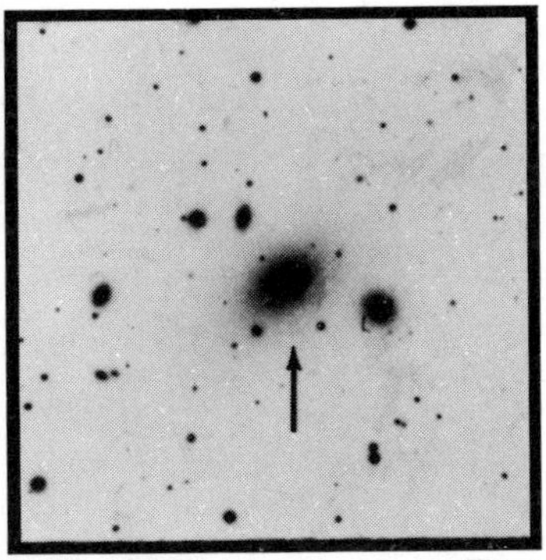

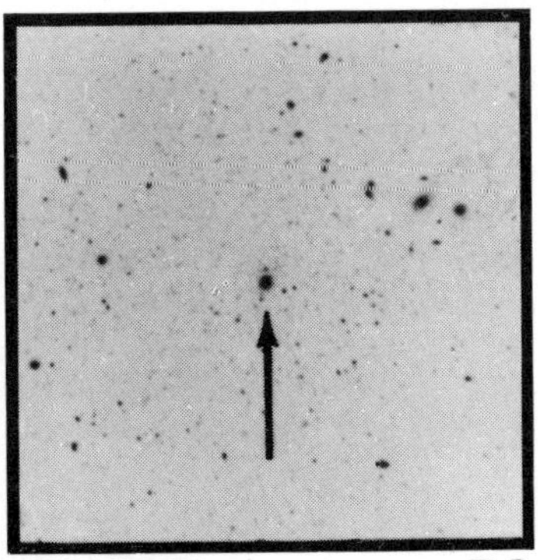

49 Various types of optical objects producing radio emission. All those illustrated in the diagram are extra-galactic. (After W. W. Morgan).

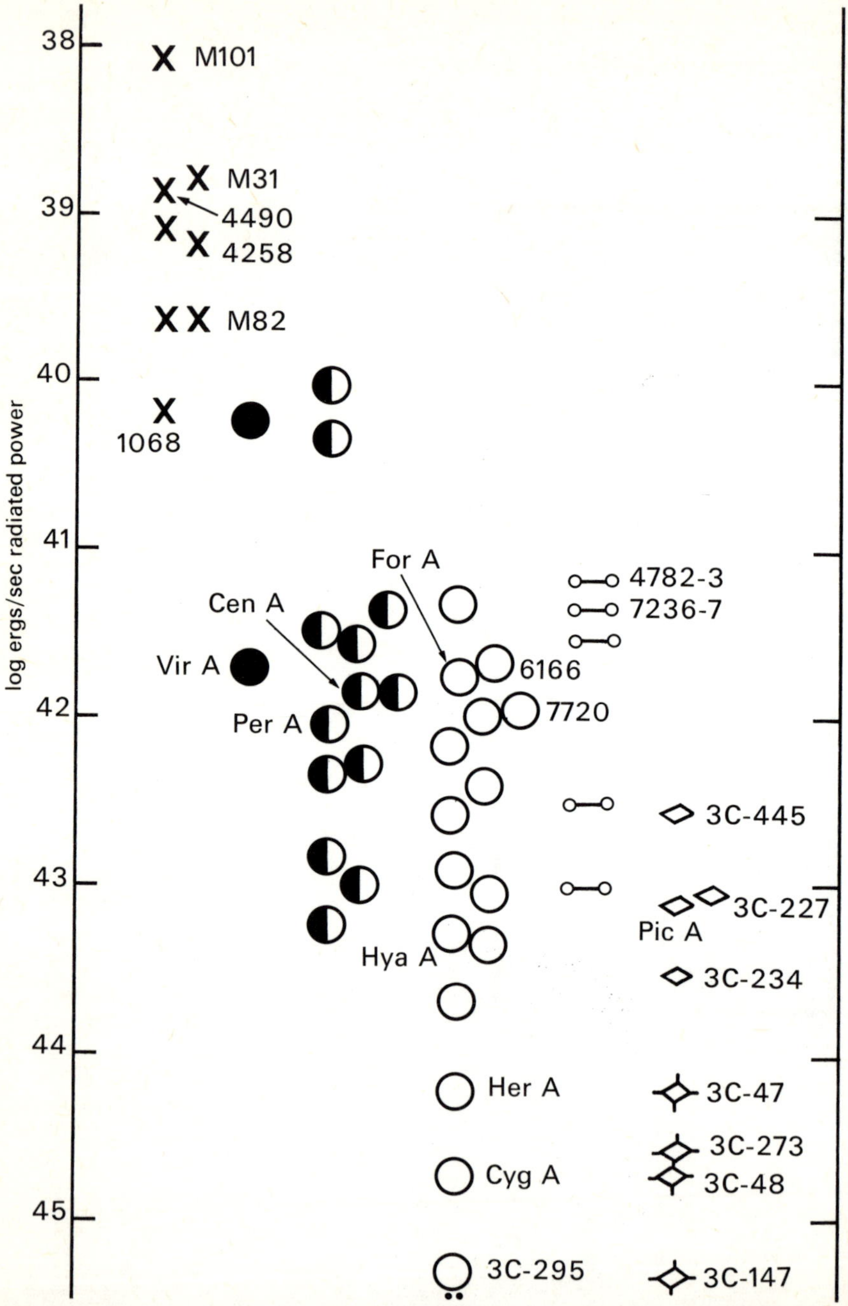

log ergs/sec radiated power
38
39
40
41
42
43
44
45
M101
M31
4490
4258
M82
1068
Vir A
Cen A
Per A
For A
6166
7720
Hya A
Her A
Cyg A
3C-295
4782-3
7236-7
3C-445
3C-227
Pic A
3C-234
3C-47
3C-273
3C-48
3C-147

50 Schematic classification of radio galaxies according to radio flux and optical type. (After Mathews and Morgan).

Spirals and Irregulars

Ellipticals (Class E)

Bright Nuclei and extended envelopes (Class D)

Intermediate between classes D and E

Brilliant, star-like nuclei and less extensive envelopes (Class N)

Quasi-stellar sources

Dumbbells (Related to D systems)

indefinitely. With a triple system, things are much more complicated, and much work remains to be done on this notorious 'three-body problem'. However, if two of the three bodies are close together and the third a long way away, the motion of the two nearby bodies will be very similar to that of two bodies, because the effects of the third may be more or less neglected. Also, the two nearby bodies may be regarded as 'one', and when considered with thc remote third body we are again back to something like the much simpler two-body problem. In this manner, a triple system may be permanent, so that its members will always stay at a finite distance from each other.

As a general rule, a third body approaching a two-body system will move in a hyperbola and will go off to infinity, after having perturbed the motions of the two bodies when at its closest to them. However, there is a slight possibility that the third body will approach in such a way that it will be captured by the two-body system. It is also possible that during the encounter, any one of the three bodies will be expelled from the system along a hyperbolic orbit. However, in most cases the final result will be that the three bodies will separate from each other, particularly when at the start of the encounter, the three bodies are roughly equidistant.

All this applies equally to quadruple systems, whose stability depends essentially on the way in which the four components

51 The radio source NGC 1275 in Perseus, photographed with the 200-in Hale reflector at Palomar. Extragalactic radio sources were once thought to be due to galaxies in collision, but this idea has now been abandoned because the collision process would not provide nearly enough energy to account for the radio emission. Unfortunately it must be admitted that the reason for this great power in the radio range, shown by the Perseus source and others, is still unexplained.

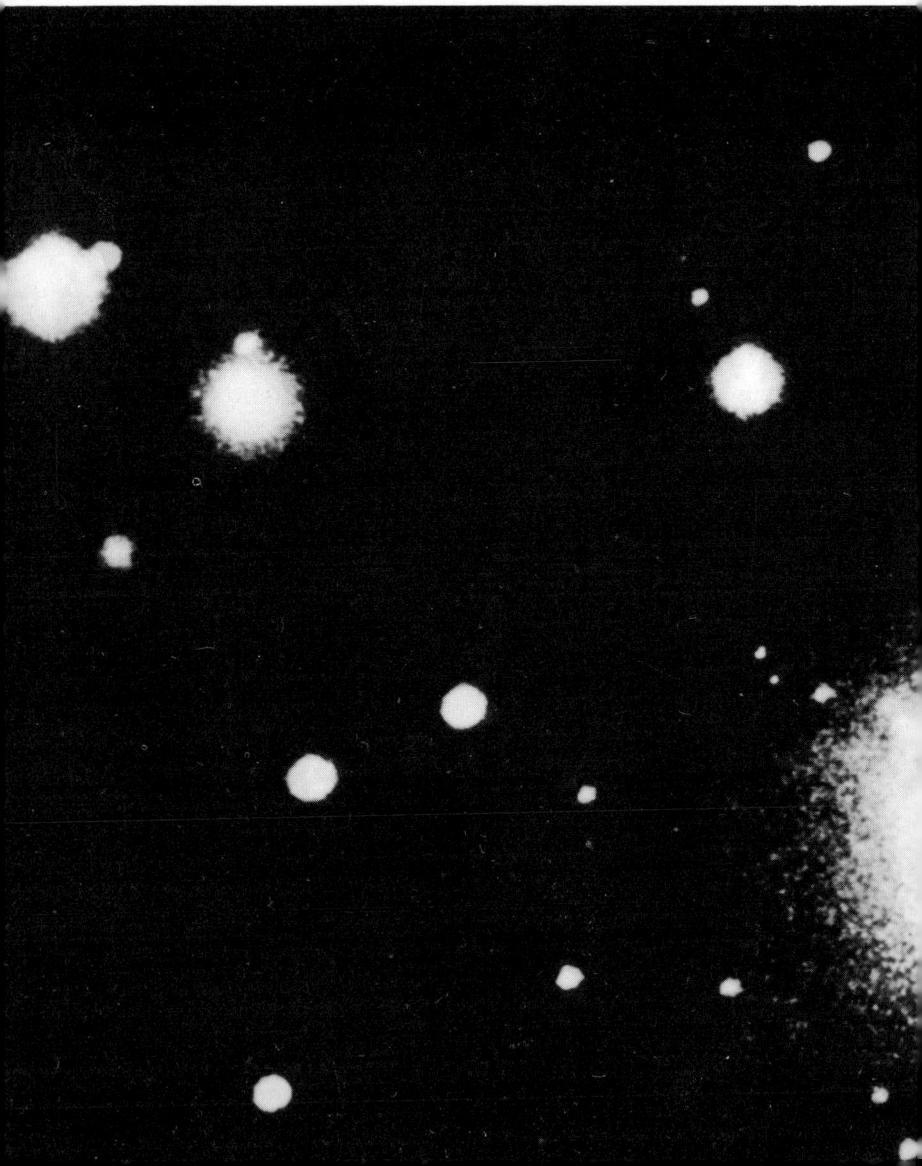

are placed in space. A system in which the four components are roughly equidistant from each other makes up a 'trapezium', and in 1950 the Armenian astronomer V. Ambartsumian of the Soviet Union showed that such systems dissociate quickly, so that the components go their separate ways.

Ambartsumian also drew attention to the properties of multiple galaxies. It is true that a double galaxy will follow much the same rules as a binary star, so that a two-galaxy system can last indefinitely, but the situation for a triple-galaxy system is different, and cannot be compared with that of a triple-star system. Some multiples are stable, but others are temporary, and are due only to the chance passing-by of independent galaxies. Many permanent systems are known, but only a few temporary ones have been observed as yet. Some temporary systems coincide with radio sources, and it used to be thought that strong radio sources might be due to galaxies in collision, but unfortunately this attractive theory has had to be given up, because the colliding process would not provide as much energy in the radio range as is actually observed.

Many types of multiple systems have been found, and it often happens that several kinds of structures can be combined in the same galaxy. In our own Galaxy there is a vast halo of Population II stars together with the interstellar gas and Population I stars of the spiral arms, so that in a way it may be said that our Galaxy is a combination of an elliptical galaxy with a spiral. In one particular system, it is possible that there may be normal spiral structures, barred spirals, and ellipticals. These different structures correspond to the coexistence, in the same system, of sub-systems in different states of motion.

Galaxies may be very close to each other, and yet clearly separate. There are, for instance, some double galaxies which seem to be revolving round each other, much as with the components of a binary star. And in other cases, two or more galaxies may be joined by faintly luminous bridges. There may be such a bridge between our Galaxy and the Large Magellanic Cloud, and there is certainly one linking the Andromeda Spiral with its barred-spiral companion NGC 205. A tightly-packed multiple system made up of five galaxies joined by bridges of material is of special interest; it is known as Stephan's Quintet, and consists of two ellipticals, two barred spirals, and one galaxy of less definite form. Systems in which the relative separations are wider (of the order of three times the diameter of each galaxy) may be linked by immense filaments of unknown nature.

When multiple systems are studied with regard to stability, it seems clear that some of the systems cannot have existed for more than 1,000 million years. This fact is difficult to explain, and it introduces the problem of the origin of the galaxies, because it suggests that galaxies are being formed in the universe at the present time.

Masses of galaxies

A galaxy is a system in rotation, with the centrifugal force and the gravitational force in equilibrium. The mass of a galaxy can therefore be worked out by a straightforward application of the law of gravitation, provided that the speed of rotation is known. The general idea is as follows:

Newton's law of universal attraction states that the force between two bodies is proportional to the product of their masses, and inversely proportional to the square of the distance between them. For the sake of simplicity, let us assume that the Earth and the other planets move round the Sun in circular orbits. It is then easy to explain Kepler's third law, according to which the squares of the periods of revolution are proportional to the cubes of the orbital radii.

To all intents and purposes, for a body which moves with uniform velocity along a circle of radius r, the centrifugal force will be proportional to the radius of the circle and inversely proportional to the square of the period. The force exerted by the central body on the moving body is inversely proportional to the square of the radius of the circle. This gives the simple proportion:

$$\text{centrifugal force} = \frac{4\pi^2 r}{P^2} = \text{force of attraction.}$$

Kepler's third law then follows:

$$\frac{P^2}{r^3} = \frac{4\pi^2}{Gm}$$

The quantity G is the constant of universal gravitation. It can be dispensed with if the period P is expressed in years, the orbital radius r in astronomical units (one astronomical unit being the radius of the Earth's orbit), and the mass m in solar units (the mass of the Sun being taken as 1). The velocity of the circular motion is given by:

$$V^2 = \frac{Gm}{r}$$

Here, too, G can be dispensed with if m is taken in solar units, the velocity is measured in kilometres per second, and the distance is given in parsecs. However, this means that a numerical coefficient must be introduced, giving:

$$V^2\ (\text{km/sec}) = 0{\cdot}0044\ \frac{(m/\odot)}{r\ (\text{parsecs})}$$

where $m/\odot$ is the mass in solar units. This, then, is a reliable way in which to determine mass, thus:

$$(m/\odot) = r\ (\text{parsecs}) \times V^2\ (\text{km/sec}) \times 230.$$

When the attracting body cannot be reduced to a central mass of very small size compared with the orbital radius, Kepler's law no longer applies and must be replaced by a more complex law worked out from the always valid law of universal gravitation.

Some idea of the mass of our own Galaxy may be found if the last formula is taken, with r (orbital radius) = 10,000 parsecs and the orbital velocity of the Sun taken as 250 kilometres per second. The

Table 11

Name of galaxy	Type	m/10 m	m*/L*	Distance	Distance modulus, m–M	Method used
Our Galaxy	Sb					
M31	Sb	450	1·7		24·6	Rotation
M32	E	1·8	13·5			Virial
M33	Sc	8	3			Rotation
M81	Sc	170	11		27·8	Rotation
NGC 55	Magell-anic	20 to 40	3 to 6		24·6	Rotation
NGC 157	Sc	60	1·9	24 Mpc*		Rotation
NGC 253	af, Sb Sc	240	12	H		Rotation
NGC 300	Sc	21	9		26·6	Rotation
NGC 720	Es5	75				Rotation
NGC 1068	gS2p	20	2·7	H		Rotation
NGC 1084	Sc	12	1	H		Rotation
NGC 1097	SBb	5 to 13		16 Mpc		Rotation
NGC 1355	SBc	22 to 34		20 Mpc		Rotation
NGC 2146	Sa	20	3	12		Rotation
NGC 2903	Sc	3·7	5·5	6 Mpc		Rotation
NGC 3115	E7	70				Rotation
NGC 3379	E	110	12			Virial
NGC 3504	SBb	9	0·7	H		Rotation
NGC 3556	Sc	8 to 12	1·5 to 2	H		Rotation
NGC 4278	E1	38	16	H		Rotation
NGC 4486	EO	1400	33	160 Mpc		Rotation
NGC 4528	Sb	120	13	H		Rotation
NGC 4594	Sb	200	3		30·8	Rotation
NGC 4531	Sd	24	6	4·4 Mpc		Rotation
NGC 5005	Sb	105	6·3	14·4 Mpc		Rotation
NGC 5055	Sb	60	3 to 4	H		Rotation
NGC 5128	DE3	200	13	5		Radio source
NGC 5248	Sc	3	4	H		Rotation
NGC 5383	SBb	40		H		Rotation
NGC 7479	SBb	16		H		Rotation
VV 254		130				Rotation

* Mpc = megaparsecs; one megaparsec = one million parsecs.

Further explanations of this table are given later.

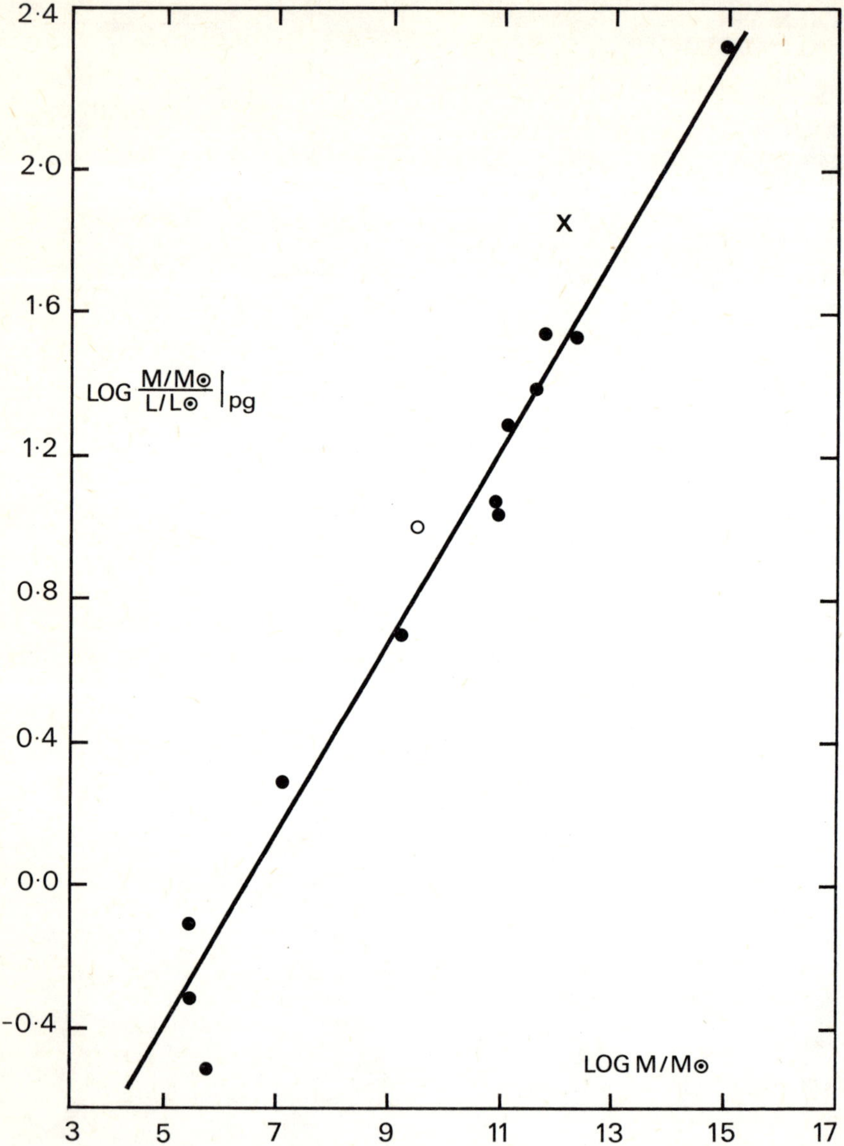
2·4
2·0
1·6
1·2
0·8
0·4
0·0
-0·4
LOG M/M⊙ / L/L⊙ |pg
X
LOG M/M⊙
3
5
7
9
11
13
15
17

52 The mass/luminosity relationship for globular clusters in galaxies and systems of galaxies. The cause of this relationship is unknown.

mass of the Galaxy comes out at something like 140,000,000,000 times that of the Sun.

If the major axis of the image of a galaxy is placed along the slit of a spectrograph, the rotational velocity can be found from the Doppler effect. Once the distance of the galaxy is known, the mass can then be calculated. More than 30 galaxies (including our own) have been studied in this way. Among the thirty listed in table 11, are 19 spirals, 5 barred spirals, 4 ellipticals, 1 D-galaxy (strong radio source) and one peculiar galaxy (a member of a double-galaxy system).

From the table, it is seen that the masses of galaxies range between several thousand million times that of the Sun up to several hundred thousand million solar masses. The ratio (m^*/L^*) between mass and intrinsic brightness, expressed in units of solar mass, differs between the spirals and the ellipticals. For spirals, the ratio lies between 0·7 and 10, with a mean value of about 4, while for the ellipticals the ratio is about 12. Mass for mass, the spirals are more luminous than the elliptical galaxies.

The values given in table 11 are not definitive, because they depend entirely upon the distance estimates, which are by no means certain, even though they are of the right order.

If distance is determined by the distance modulus, the calculated mass is proportional to the distance and the luminosity is proportional to the square of the distance; the ratio mass/luminosity is proportional to the inverse square of the distance, as given by $(m/L)\alpha r^{-2}$. If distance is determined by means of the Hubble-Humason law, the calculated mass is inversely proportional to Hubble's constant and the luminosity is proportional to the inverse square of that constant, while the velocity of recession may be expressed as V (recession) = Hr. This gives $m(1/H)$, $L\ (1/H^2)$, and $(m/L)H$. H, of course, is Hubble's constant.

The mass/luminosity relation

Systems which are poor in interstellar absorbing matter make up what may be regarded as a homogeneous group. The late Walter Baade classified stars in these systems as belonging to Population II. It is notable that the systems include an abnormally large number of faint blue stars.

It is remarkable to find a relation between the colours of these Population II systems and their intrinsic brightness, with the bluer systems being the less luminous. Moreover, in examining the sequence of objects starting with globular clusters (M92 and M3, for example), and continued through elliptical galaxies (such as M32 and NGC 3379) ending with the galaxies in the Coma cluster, to be described below, it is found that though luminosity increases regularly according to mass, it does so less rapidly than might be expected. The relation between mass and luminosity is as follows:

$$(L/L\odot) = (m/3{\cdot}10^5\ m\odot)^{0{\cdot}73}\ .\ 3 \times 10^5$$

Another notable feature is the regular decrease in the mean density of the systems from the globular clusters through to the cluster of galaxies in Coma. Probably this is associated with a regular change in the conditions under which stars are formed, the masses of the individual newly-formed stars increasing while their density becomes less. All this indicates that in the Coma cluster there must still be great quantities of gas which have not condensed into galaxies. (More about this will be said later.) Also, the existence of this relationship casts some doubt on our methods of finding the mean density of matter in the universe, which is an important problem in cosmology.

The formation of spiral galaxies

In our Galaxy, the older objects move out to great distances from the main plane, while the younger objects tend to lie much closer to the plane. The flattening of the original star-forming material toward the equatorial plane must have taken place so quickly that it may be regarded as a collapse process. It is important to find out whether the same sort of thing has happened in other spirals, and fortunately this can be done, by virtue of the important mechanical property of matter known as the *conservation of angular momentum*.

Momentum is the produce of mass and velocity. With circular motion, the product of the amount of movement and the distance from the axis of rotation is known as the *angular momentum*. If there are no external forces to be considered, and the mass is constant, the angular momentum must also be constant; in other words, angular momentum is constant in an isolated system. Imagine, for example, a body which may be regarded as a point, attached to a light string and moving in a circle with uniform velocity. If the length of the string is halved, the speed of the moving body will be doubled; if the length of the string is doubled, the velocity is halved.

Studying the movements in a spiral galaxy make it possible to calculate the value of angular momentum contained in a cylindrical ring of radius r and thickness of one centimetre. We are, in fact, considering star-forming material which is originally diffused through a spheroidal volume, and rotating with uniform circular movement; and this sounds reasonable enough, since star-forming material is gaseous and subject to violent turbulence. The viscosity of the turbulent motion will be sufficient to make the rotation become uniform, as would happen in the case of a solid. And if the nebulosity is not in equilibrium, there is nothing to prevent its collapse toward the equatorial plane.

The amount of angular momentum contained in a nebulous cloud in the form of a cylindrical ring of radius r and one centimetre thickness can be calculated, and the measured angular momentum in a galaxy can be compared with the calculated angular momentum in a nebulosity which is rotating uniformly. With seven spirals, it has been found that the two quantities are equal, within the limits of errors of measurement. As with our Galaxy, it seems then that angular momentum is conserved during the collapse of a nebulous cloud into a spiral shape.

Spiral structure

Observations of spiral galaxies show that objects of different kinds contribute to the spiral arms. Photography makes it clear that these objects are luminous blue stars together with dark and bright nebulosities. In our Galaxy and the Andromeda Spiral, radio studies at a wavelength of 21 centimetres have led to the discovery of the positions of neutral hydrogen clouds in the spiral arms. In our Galaxy, in the neighbourhood of the Sun, it seems that the arms are trailing with respect to the general rotation of the system.

The fact that spiral galaxies are flatter than ellipticals has long lent support to the view that the differences between the two types are due mainly to the different speeds of rotation. Yet this cannot be the whole story, because elliptical galaxies contain little of the dust and gas which is so abundant in the barred and normal spirals.

The formation of spiral arms is certainly linked with the distribution of the nebulous material in space, because the very luminous blue stars characteristic of the arms cannot be more than ten million years old. This must be compared with the time taken for the stars to be dispersed through the galaxy and with the time during which the arms have persisted; there seems no doubt that the internal movements in the arms mean that the spiral arms themselves must last for something like one hundred million years.

Though the origin of the spiral arms is undoubtedly associated with the properties of the motion of nebulous material inside the galaxies, there are difficult problems to be raised concerning the way in which gas condenses by virtue of gravitation, the origins of the stars (which then no longer share the motions of the clouds), and the magnetic field. The whole question is so complex that at the moment we must confess that we do not understand it, and as yet there is no satisfactory theory by which the different forms of spirals can be classified. All that can be done is to use the straightforward classification given at the start of the present chapter, and to continue the search for correlations and relationships between theory and observation.

Elliptical galaxies

Unlike the spirals, the elliptical galaxies have small angular momentum, which explains why they are not markedly flattened. And while the essential feature of spirals is their overall rotation, the general movement of an elliptical galaxy is slight; the essential characteristic is the movement of the stars in all directions, rather in the fashion of the molecules in a gas. There are strong analogies between studies of elliptical galaxies and of globular clusters. The same fundamental principles of mechanics are applicable to each.

The virial

The sum of the potential energy and twice the kinetic energy of a system is called the *virial*. With a system in equilibrium, the virial is zero. The simplest case involves a double system in which the movement is circular and uniform, so that the distance between the two components never changes. The virial is zero because the centrifugal force is equal to the force of attraction. And since a system containing a large number of stars may be regarded as being in a state of equilibrium, its virial, too, is zero.

When the radial velocities of the objects in a system are measured, the kinetic energy per unit mass can be calculated. The kinetic energy of the whole system is equal to the product of its mass by the kinetic energy per unit mass. Moreover, the potential energy is proportional to the square of the mass and inversely proportional to the radius of the system. The radius can be measured, and by applying the virial theorem the mass of the system can be found.

This method has been used for three galaxies; Messier 32, NGC 3379, and the nucleus of Messier 31 (the Andromeda Spiral). The mean velocity of the stars was determined by measurements of the spectra of the galaxies, because the individual motions of the star cause broadening of the spectral lines due to combined Doppler effects.

Table 12 Relaxation times for clusters and elliptical galaxies

Mass	**Radius (parsecs)**	**Relaxation time** (thousands of millions of years)
10^6	100	40
10^9	1,000	200,000
10^{12}	10,000	160,000,000

Encounters between stars

The stars in a cluster behave rather like the molecules of a gas. There are movements in all directions, with each star going its own way, but because the stars are so widely spread out they do not interfere with each other. However, over a sufficiently long period of time, each star will have its path modified until the whole collection of stars is completely mixed; no trace will be left of the situation as it used to be when the system came into existence. The time required for mixing is called the *relaxation time* of the system. It increases with the number of stars concerned and with the size of the system.

While the stars can be regarded as moving independently of each other, the calculated relaxation times are relatively long. For stars with masses equal to that of the Sun, the relaxation times are as in table 12. These values mean that elliptical galaxies can never attain statistical equilibrium, simply because the individual stars take so long to influence each other. The great length of the relaxation time does not necessarily mean that the mass values obtained by use of the virial theorem are dubious, because even though the virial for a stationary system is nil, the theorem gives no information about the velocities of the separate stars.

Table 13 Masses of clusters of galaxies and of individual galaxies in clusters

Cluster	Mean radial velocity	Number of galaxies	Mass of cluster	Mass of galaxies	m*/L*
Hercules	631	73	75,000	1,000	120
Virgo	643	500	1,000,000	2,000	600
Coma	1,010	670	1,000,000	1,600	370
Perseus	437	360	1,000,000	770	230
Cancer	427	84	1,000,000	830	300

And even if statistical equilibrium is not attained, the system could still be in a static state, which alone is enough to justify applying the theorem.

Statistical equilibrium can be attained even if the stars take part in collected or co-ordinated movement instead of moving independently. Moreover, a small number of very massive objects will reach equilibrium more quickly than a large number of objects with small mass occupying the same volume of space. A group of stars moving in a co-ordinated manner behaves like a single object. In an elliptical galaxy, then, it is possible that the co-ordinated movements of the stars result in a static state being reached comparatively quickly.

Energy of elliptical galaxies

If the mass/luminosity relationship is taken as being constant for all elliptical galaxies, it is possible to calculate the gravitational energy and then work out the relation between this energy and the mass of the system. There is no need to take the gravitational energies of individual stars into account; we are dealing with the energy which keeps the stars together in a system. It is found that the gravitational energy increases steadily as the 3/2 power of the mass. Let us compare, for instance, a galaxy with a mass 10,000 million times that of the Sun, which has a gravitational energy 100,000 times that of the Sun, with a galaxy with a mass a million million times greater than the Sun's, which has a gravitational energy 100 million times greater than that of the Sun.

If the mass/luminosity relation for elliptical galaxies is accepted as valid, then the gravitational energy holding the stars together in the system will increase as the 1·63 power of the mass. This result is in good agreement with the law described above.

The increase of what may be called 'binding energy' with mass can be explained on the assumption that elliptical galaxies were formed from a hot, contracting cloud of gas. The stars would be formed later, inside the proto-galaxy, by condensation of the gas. Presumably the shrinking of the proto-galactic gas came to an end when radiation could no longer escape from it. The time when radiation begins to be trapped gives a relationship between the radius of the galaxy and its density, and consequently between the radius and the total mass. The calculated 'binding energy' agrees remarkably well with the actual binding energy of stars in elliptical galaxies.

And yet the problem of how stars were formed inside the mass of the proto-galactic material is not solved in this way. So far as we know, it seems that the originally hot, turbulent gas must have cooled down because of the random formation of denser regions, so that stars could start to form in the cooler regions. The density distribution in a present-day galaxy is, to a certain extent, an echo of the density distribution in the turbulent gas of the proto-galaxy.

Clusters of galaxies

Fifteen large clusters of galaxies have been studied with respect to the distribution of individual galaxies within the cluster. Three clusters, those in Hercules, Virgo and Coma Berenices, have been studied in detail, and the velocities of the separate galaxies inside the clusters have also been investigated.

The regularity of the distribution of galaxies in these clusters suggests a condition of equilibrium, and attempts have been made to apply the virial theorem so as to obtain the mean value of the masses of the galaxies in the clusters. The results are given in table 13. It will be noticed from table 13 that the value of the mass/luminosity relationship is much higher for the clusters of galaxies than it is for measures of the masses of individual galaxies. There are three possible explanations for this:

a The galaxies in a cluster are much more massive than isolated galaxies.
b The galaxies in a cluster are of the same mass as isolated galaxies, but the cluster is rich in intergalactic material which is completely or virtually invisible.
c The clusters are not in equilibrium, so that the properties of the virial theorem do not apply.

The first explanation (a) can be rejected out of hand, because there is nothing whatsoever to distinguish a galaxy in a cluster from a solitary galaxy, but the second idea (b) has more to recommend it. In the first place, in a rich cluster of galaxies, such as those in Coma, Perseus or Cancer, the distribution of the individual galaxies in space follows the same familiar pattern as the molecules of a gas in equilibrium in its own gravitational field. If the velocities of the member galaxies do not result from the effects of encounters between themselves, but are due to the original state of affairs, then the establishment of equilibrium will be roughly equal to the time which an object would take to cross the system. For the clusters in Coma Perseus and Cancer, the relevant periods are of the order of 20,000 million years, 50,000 million years and 40,000 million years. This means that the Coma cluster may be in equilibrium, and also perhaps the Perseus and Cancer clusters.

In the Coma cluster, there is a distinct segregation between the faint and brighter galaxies. The bright galaxies, of greater mass, are grouped together in the central part of the cluster, while the fainter and less massive galaxies are proportionately more numerous at the periphery – just as a light gas in equilibrium in a gravitational field will diffuse further than a heavy gas. An arrangement of this kind is a clear argument in favour of a state of equilibrium.

The next step is to explain the difference between the mass as observed in the galaxies and the mass calculated by applying the properties of the virial. The straightforward answer is to suppose that the difference is due to the presence of intergalactic matter. The eminent astrophysicist F. Zwicky has shown that the Coma cluster acts as an obscuring region, hiding clusters of galaxies which are further away from us. The existence of a mass/luminosity relationship for elliptical systems which contain no dust, valid for globular clusters as well as the cluster of galaxies in Coma, seems also to favour the idea that there is a great deal of invisible nebular material.

The third possibility (c), due to V. Ambartsumian of the Soviet Union, is that the clusters of galaxies are not in a state of equilibrium. This idea is based on a comparison with the double and multiple galaxies. It is possible to derive masses for the double galaxies on theoretical grounds, and the values obtained are in good agreement with the values found by studies of the rotations of galaxies. However, multiple systems with three or four component galaxies give systematically higher mass values than those derived for the double galaxies. From an analogy with multiple stars, which are known to be unstable systems, Ambartsumian has suggested that multiple galaxies also are unstable systems in the process of dissociation. In this case, the observed differences in velocity will be greater than those obtaining in a state of equilibrium, and will correspond to escape velocities of one or several members of the system.

Ambartsumian goes on to suggest that the clusters of galaxies are themselves in the course of dissociation, and that the observed differences in velocity of the separate galaxies are greater than would be the case for a system in equilibrium, so that they are nothing more than the velocities of dispersion of the clusters into space.

Quasi-stellar radio sources

The interpretation of quasars or quasi-stellar radio sources (Qs) is a very difficult matter, and as yet our knowledge is very incomplete. It has been said, with truth, that the quasars are the most remarkable objects ever discovered. They are often termed Qs galaxies, but we cannot be at all sure that they are galaxies in the accepted sense of the term.

A quasar may radiate some 2,500,000 million times more powerfully than the Sun; its lifetime, calculated from the size of the radio source, must be of the order of 100,000 years to 1,000,000 years, so that during its career it radiates something like 10^{52} joules. Most of this energy is synchrotron radiation, from high-energy electrons which quickly lose this energy simply because they are radiating. Therefore, the source must contain high-energy protons (that is, cosmic-ray protons) whose energies are at least one hundred times greater than those of the electrons. The protons themselves have been accelerated by a mechanism whose efficiency, judged from studies of cosmic rays, is not greater than 1/100.

This tremendous quantity of radiation cannot be explained unless it is possible to discover a source which is capable of providing 10^{56} joules, and this is no easy matter. It seems that one solution is to suppose that the energy is set free by the gravitational collapse of an enormous mass. The continuous radiation of a quasar such as 3C-48 or 3C-247 appears to come from a mass with a radius of about one parsec; such an object would have the gravitational energy needed, and the mass itself would be of the order of 100 thousand million times that of the Sun. Numbers of this sort are so far beyond our everyday experience that they are quite impossible to appreciate.

Indeed, the quasars seem to be so incredible that some astronomers have tried to explain them in other ways. One plausible answer is that the quasars are not extragalactic at all, but are objects belonging to the halo of our Galaxy; in this case they would presumably have been ejected from the galactic centre with high velocity.

53 The ring nebula M57 in Lyra, photographed with the 200-in Hale reflector at Palomar. This is the best example of a planetary nebula – that is, a faint star surrounded by an extensive gaseous shell which has been found to be expanding. It has been suggested that a planetary nebula may result from the continuous ejection of matter during a certain phase of the evolution of the central star. M57 is visible with a small telescope, but the central star, as in all planetaries, is very faint.

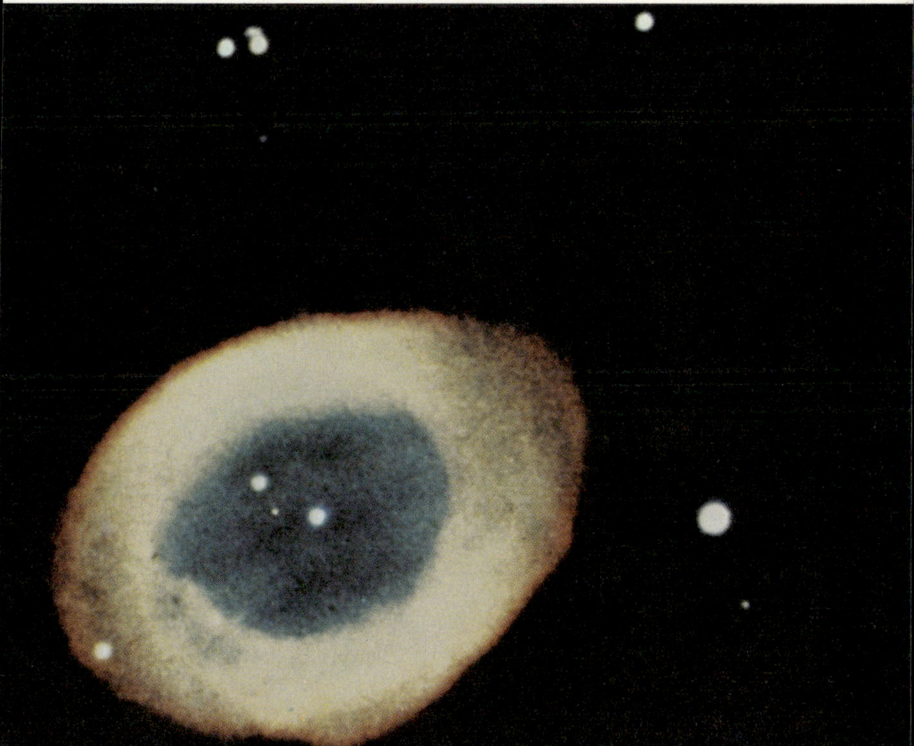

According to this theory, the quasi-stellar sources would be only about 25,000 times as luminous as the Sun, and their kinetic energy would be of the order of 10^{49} joules. And yet the energy of such an explosion, which would be supposed to have taken place in the centre of our Galaxy some hundreds of thousands of years ago, would be too low, by at least one order of magnitude, to account for the observed number of quasars. On the whole, it seems more likely that a quasar is a remarkable kind of elliptical galaxy at a tremendous distance from us, but the last word has by no means been said.

Conclusion

At this point it may be as well to sum up what is known about the galaxies, since this is an essential preliminary to an overall study of the universe.

The realm of the galaxies is more complex than that of the stars; we have begun to work out the mechanism of stellar evolution, but our ideas about the evolution of the galaxies is at best rudimentary, and could even be called nonexistent. It is all very well to classify galaxies by means of their forms, but it is important not to underestimate the difficulty of classifying objects which show such amazing variety. Though Hubble's simple division of galaxies into elliptical, spiral and irregular systems is still useful in its way, it is not detailed enough for full analysis. Attempts have been made to establish intermediate stages between ellipticals and spirals, spirals and barred spirals, and spirals and irregulars, but the correlation with spectral properties is poor, and it may be wiser to complete Hubble's classification by adding new categories such as diffuse galaxies, nebular galaxies and quasi-stellar radio sources. In this scheme, the very luminous diffuse galaxies might fill the gap between the ellipticals and the quasi-stellar sources. The quasars are at least a hundred times more luminous than normal galaxies, and certainly represent the most astonishing discovery of the last few years. Because they are so luminous, they can be observed out to immense distances, and allow us to probe further into space than we could otherwise do.

Studies of individual galaxies and their movements make it possible to find out something about the masses of galaxies. Remarkably, the mass/luminosity relationship between galaxies is more or less constant between one object and another. Taking the Sun as being of unit mass and unit luminosity, the mass/luminosity relationship for galaxies ranges between 1 and 20. This gives a key to the luminosities of the galaxies, and hence to the average density of matter in the universe.

Unfortunately, the results obtained in this way cannot be taken as wholly reliable. So far as double galaxies are concerned, the motions are not unlike those of binary stars, and lead to values for the masses which agree quite well with the figures obtained by the studies of the internal motions, but the agreement is very poor with the multiple galaxies, where the derived values are much higher. The result may be expressed as follows: If a system is in a state of equilibrium, there is a balance between the potential energy and the kinetic energy of the system. The kinetic energy depends on the masses of the components and the square of their velocity, so that if the mutual distances and velocities are known it is possible to work out the relationship between potential and kinetic energy, which in turn gives the mass. But by comparison with the masses of galaxies as found by the rotation method, the relative velocities of the individual members of a multiple galaxy are too high for the system to be in equilibrium. Moreover, the masses as calculated by the observed velocities are appreciably greater than the masses as calculated from studies of internal motions. For clusters of galaxies, the situation is even worse, because if the systems are in equilibrium the masses of the individual galaxies would have to be enormously greater than their luminosities indicate.

It is all very puzzling, and the answers have not yet been found. Either these immense systems are not in a state of equilibrium, in which case they must be about a thousand million years old (an

54 The great Andromeda Spiral, M31 (NGC 224), photographed in colour with the 48-in Schmidt telescope at Palomar. This spectacular galaxy is only 2,200,000 light-years from the earth so that it can be examined in detail. It has been found to be considerably larger than our Galaxy and so is the senior member of the Local Group.

55 Spiral galaxy NGC 7331 in Pegasus, photographed with the 200-in Hale reflector at Palomar. The spiral structure is well shown, but is very difficult to see visually because the galaxy is much further away than the Andromeda Spiral and members of our Local Group. It is, however, typical of spirals of this kind.

idea which involves continuous formation of galaxies throughout the universe), or else the very high values calculated for the masses are due to significant amounts of invisible material scattered between the galaxies in a cluster. This latter idea sounds plausible enough in view of the mass/luminosity relationship, but it must be added that the significance of this relationship is not yet clear, and it may be that, after all, the clusters of galaxies are not in equilibrium.

Little is known about the evolution of the galaxies. Undoubtedly our Galaxy was produced as a result of a gravitational collapse of its material toward the main plane, and the distribution of velocities in the elliptical galaxies seems to show that the same sort of thing has happened there also, but we cannot work out a true evolutionary sequence; we do not know whether a spiral evolves into an elliptical, or *vice versa*, or whether neither process is valid. However, the radio sources show that evolution in the galaxies must take place on a grand scale. In some galaxies there may be a succession of explosions, causing large quantities of material to be ejected to great distances, and so emitting radio waves for periods of hundreds of thousands of years.

In short, enormous masses of the order of one to a hundred thousand million times that of the Sun, the tremendous complexity of the galaxies, incontestable proof of some evolutionary sequence, and perhaps information about the formation of the universe itself – all these are to be found in our studies of the great star-systems which we can see in the depths of space.

5 The universe

In the previous parts of this book, we have taken a look at the various kinds of bodies which are to be found in the universe. The most important of these are, of course, the galaxies, which may be single, double, multiple, or grouped into huge clusters. Probably there are clouds of tenuous intergalactic matter spread between the galaxies, and the whole universe is permeated by high-energy cosmic-ray particles (over 10^{18} electron-volts), photons and neutrinos.

The overall study of the universe leads on to studies of the distribution of the galaxies in space, the mean density of material in the universe, and relationships between the matter which makes up the galaxies and the more subtle forms of matter such as photons and neutrinos. The first step must be to construct some kind of map of the nearby galaxies.

The structure of the inner metagalaxy

In 1933, Shapley and Ames, in America, published a catalogue of all galaxies brighter than the 13th magnitude. The catalogue included 1,249 objects, and a map could be compiled without difficulty; the result is given here. There are various interesting features of the map, notably the Virgo cluster; the groupings in Coma, Canes Venatici and Ursa Major, which lie close together in a band in the northern hemisphere; the southward extension of the Virgo cluster in the direction of Centaurus; the cluster in Leo, and the groups in the southern hemisphere which seem to form two belts of galaxies. The southern hemisphere is markedly deficient in galaxies as compared with the northern half of the sky, and there are practically no galaxies around galactic longitude 0°.

It is interesting to compare the distribution of the galaxies brighter than magnitude 13 with the distribution of those galaxies with radial velocity less than 1,500 km/sec. The diagram given here is drawn from the catalogue of Humason, Mayall and Sandage; it is very like the map described above, particularly in the northern hemisphere.

56 The distribution of galaxies whose radial velocities are less than 1,500 km/sec (after Van Albade). The sky distribution of these galaxies in the southern hemisphere is similar to that for bright galaxies, but the northern hemisphere distribution is different.

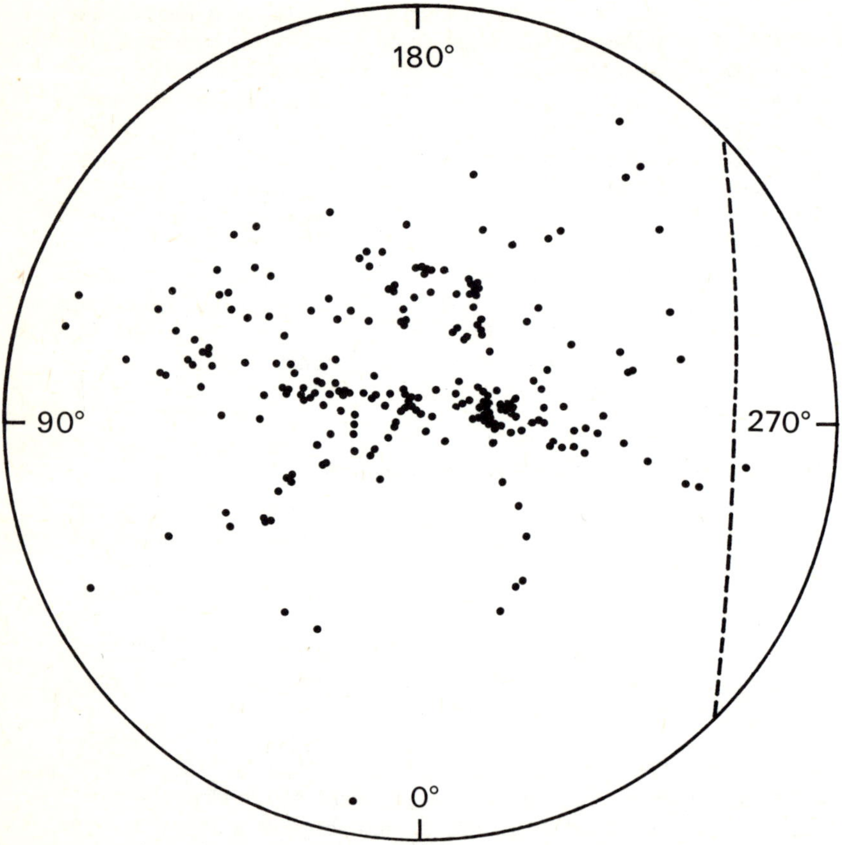

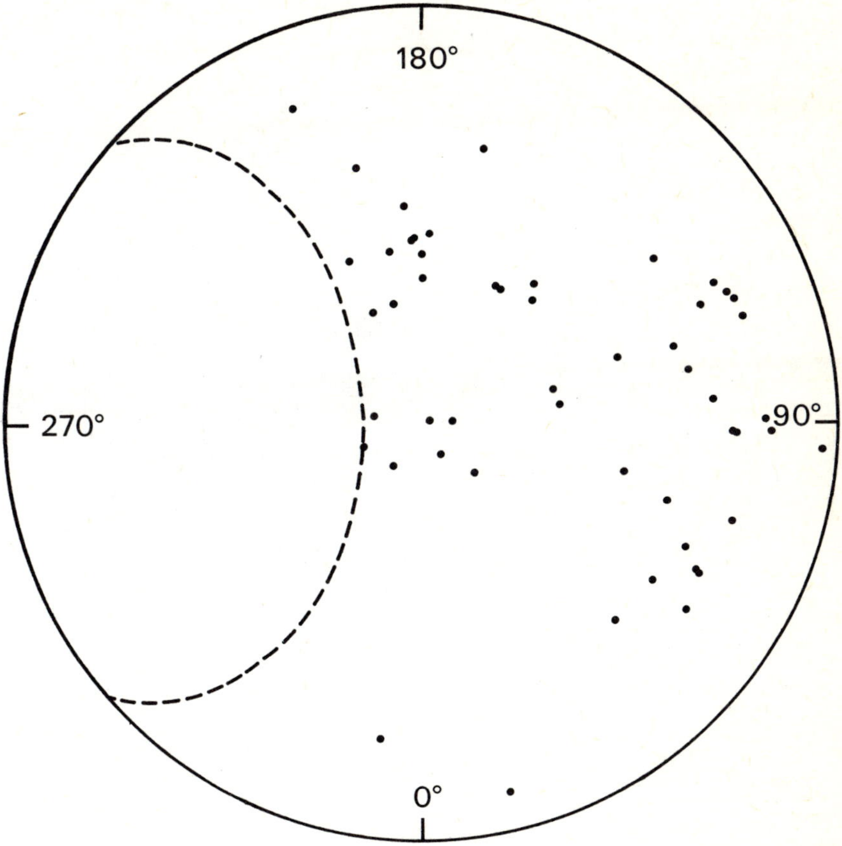
180°
270°
90°
0°

57 Galaxies brighter than magnitude 13 are not uniformly distributed over the sky (after Shapley and Ames).

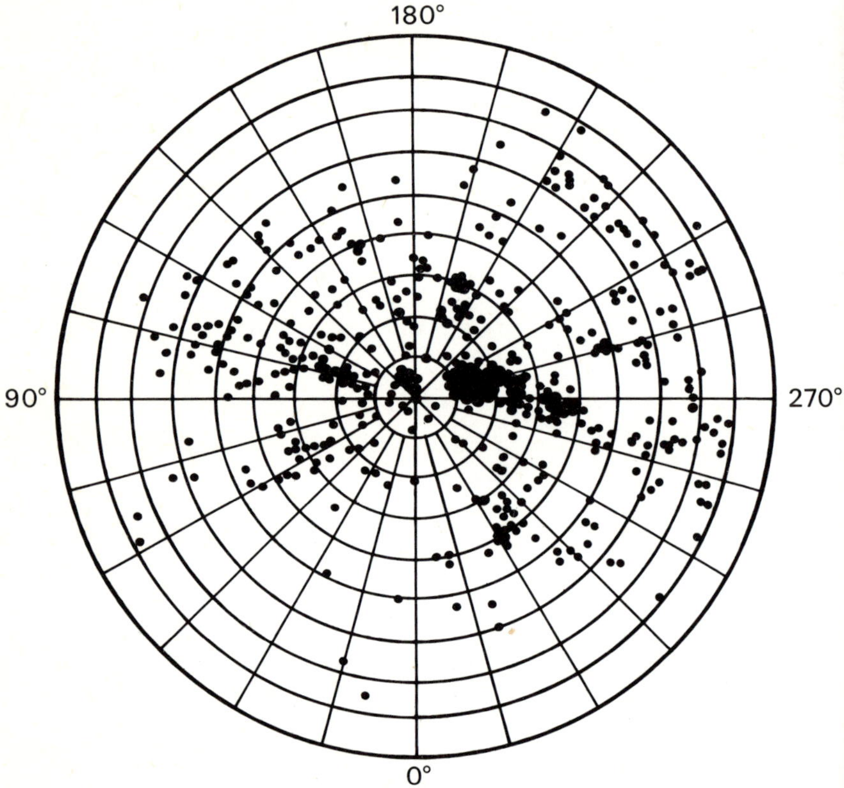

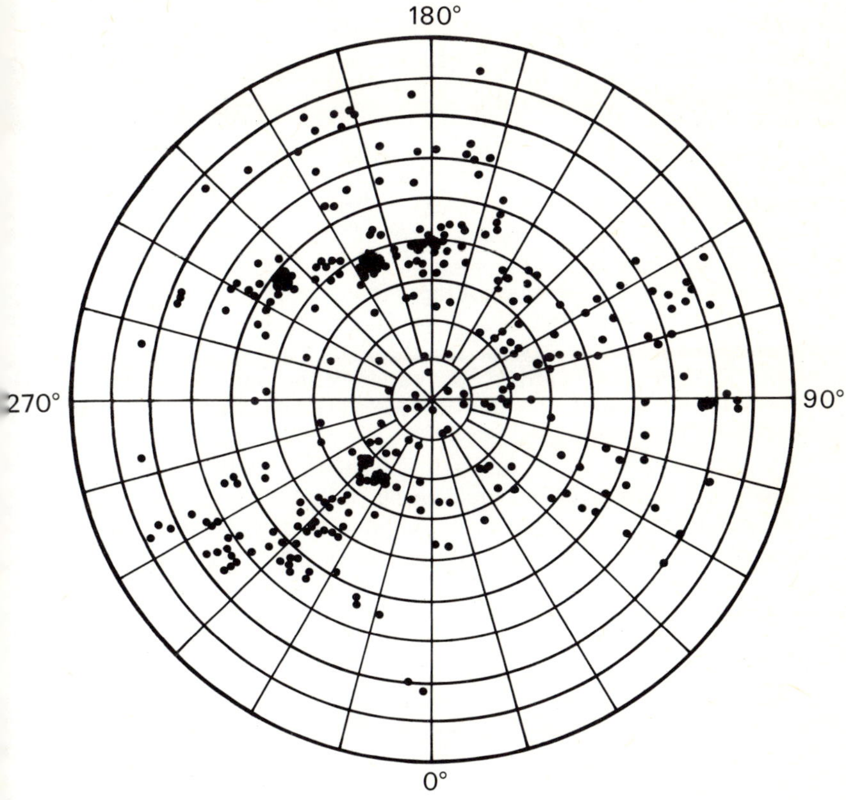
180°
90°
0°
270°

The general impression is of real grouping of galaxies in space, with sub-groups occurring inside the main system, and this sort of arrangement was studied by the American astronomer Harlow Shapley, who coined the term *supergalaxy*. In a supergalaxy, the individual galaxies are grouped in a disc-shaped system with a diameter of 15,000,000 parsecs. There is evidence of this in the night sky; the disc is marked by a belt of bright galaxies passing close to the poles of our Galaxy. There is a second belt in the southern hemisphere, passing through the constellation Fornax, which might be a second supergalaxy seen from the side. The sparse population of the southern part of the supergalaxy may be due to the fact that our own Galaxy is well away from the centre of the whole system, so that our view is unsymmetrical. And according to Gérard de Vaucouleurs, the distribution of radial velocities indicates that the supergalaxy is expanding as well as rotating differentially.

Doubts remain, however, as to the reality of this supergalaxy. According to Shapley, it would be surprising if a group of about a thousand galaxies, spread over a volume of space 30,000,000 light-years in diameter, were found to make up a coherent system. Moreover, the radio sources do not seem to have the same kind of distribution as the galaxies of the Shapley-Ames catalogue, and it must be admitted that studies in the radio range give no support to the idea of a supergalaxy.

In addition, the time needed to set up co-ordinated motion inside so fast a volume of space is in the order of 10,000 million years and, as will be shown later, this is much the same as the period during which the expansion of the universe has been going on. We seem to meet with difficulties here. It is hard to believe that so immense a system could have been formed so that co-ordinated velocities were present from the very beginning; there would be no good reason for anything of the kind. A local expansion or contraction here and there would be quite on the cards, but this is very different from a rotation, involving laws of conservation or movement which could hardly be satisfied. Yet the time which an originally un-coordinated system would take to become co-ordinated seems incompatible with the time scale of the universe as a whole. To sum up: the supergalaxy idea is attractive, and the observations of clusters of galaxies give some support to it, but up to the present time it has certainly not been proved.

More remote regions

Our knowledge of the more distant parts of the universe has grown rapidly during the last few years, and special mention should be made of the catalogue published by F. Zwicky, including all galaxies down to the 15th magnitude. But studies of still fainter galaxies can only be statistical, and depend chiefly upon actual counts. Under the direction of the American astronomer Shane, the Lick Observatory undertook to count all the galaxies down to magnitude 18·4; great care was taken to make the count complete. For fainter galaxies, down to magnitude 21, the only counts are those made years ago by Hubble, which cover very small areas, and which are of doubtful quality.

Some idea of the enormity of the task is seen by looking at Hubble's relationship between the magnitudes and numbers of galaxies:

$$\log N = 0{\cdot}6\,(m - \Delta m) - 4{\cdot}43$$

where m is a correction which depends on the red shift in the spectrum. Table 14 gives Hubble's estimated numbers of galaxies observable in the whole sky, down to various value of magnitude.

Bearing in mind that there are a thousand million seconds in 30 years, and that it takes more than a second to identify a galaxy on a photographic plate and estimate its magnitude, it is easy to see that even the Lick Observatory project of counting galaxies down to magnitude 18·4 is a gigantic undertaking. The examination of plates bearing images of galaxies down to magnitude 21 can be carried out only by automatic methods, and these have yet to be developed.

Table 14 Numbers of galaxies according to Hubble

Limiting magnitude	Number of galaxies, N (m)
18	1,650,000
19	5,600,000
20	19,000,000
21	57,000,000
22	200,000,000
23	400,000,000

Though it is clearly out of the question to make exhaustive counts, much could be learned from sample counts made in selected regions. This was Hubble's method, but unfortunately his results are now out of date, and no revision has yet been attempted.

Clusters of galaxies

The grouping of galaxies is a very general phenomenon, and ranges from multiple galaxies through to the great clusters. The tendency is well shown on the Lick maps given here. The density curves, depending upon the numbers of galaxies per square degree, show that tremendous groups occur. However, there are various difficulties about giving a full analysis, because there are several different points to be borne in mind:

1 Galaxies really do occur in groups, and the number of members may range from a mere two up to many thousands.

2 Within each group there may be galaxies of all kinds; there are relatively few brilliant galaxies, and a relatively large number of faint ones.

3 The sizes of the clusters of galaxies are not uniform.

4 Clusters of galaxies are scattered all through space, and as seen from Earth some of the clusters may lie behind others.

There are two principal ways of showing that clusters of galaxies exist. First, some of them are clearly identifiable on photographic plates. Secondly, the physical characteristics of the clusters may be obtained by detailed statistical analyses of their distribution in the sky.

In 1936, Zwicky began a systematic study of the clusters of galaxies, and his work resulted in many new discoveries. Shapley's first catalogue had contained 35 clusters, and had been followed by Abell's catalogue of dense clusters, containing 2,712 entries. Zwicky's catalogue included several thousands of clusters, and he then started to compile maps. He classified the clusters by means of their measured velocities of recession, as follows:

VELOCITY OF RECESSION, km/sec	CLASSIFICATION	
0—15,000	Near	(N)
15,000—30,000	Moderately distant	(MD)
30,000—45,000	Distant	(D)
45,000—60,000	Very distant	(VD)
60,000 and greater	Extremely distant	(ED)

The diagram shows Zwicky's maps for these five classes, in a region in Virgo close to the galactic pole. Zwicky himself noted that:

1 In regions where galactic and intergalactic absorption is negligible, VD and ED clusters are distributed in a fashion which is remarkably uniform and random, bearing in mind the selective effect due to galactic absorption when the line of sight passes close to the galactic plane and so meets with interstellar material. From this, Zwicky concluded that there was no evidence that the clusters were themselves grouped into *clusters of clusters*. Zwicky's views will be discussed later.

2 In every region containing dense N clusters, the number of VD and ED clusters is lower than might be expected. Zwicky concluded

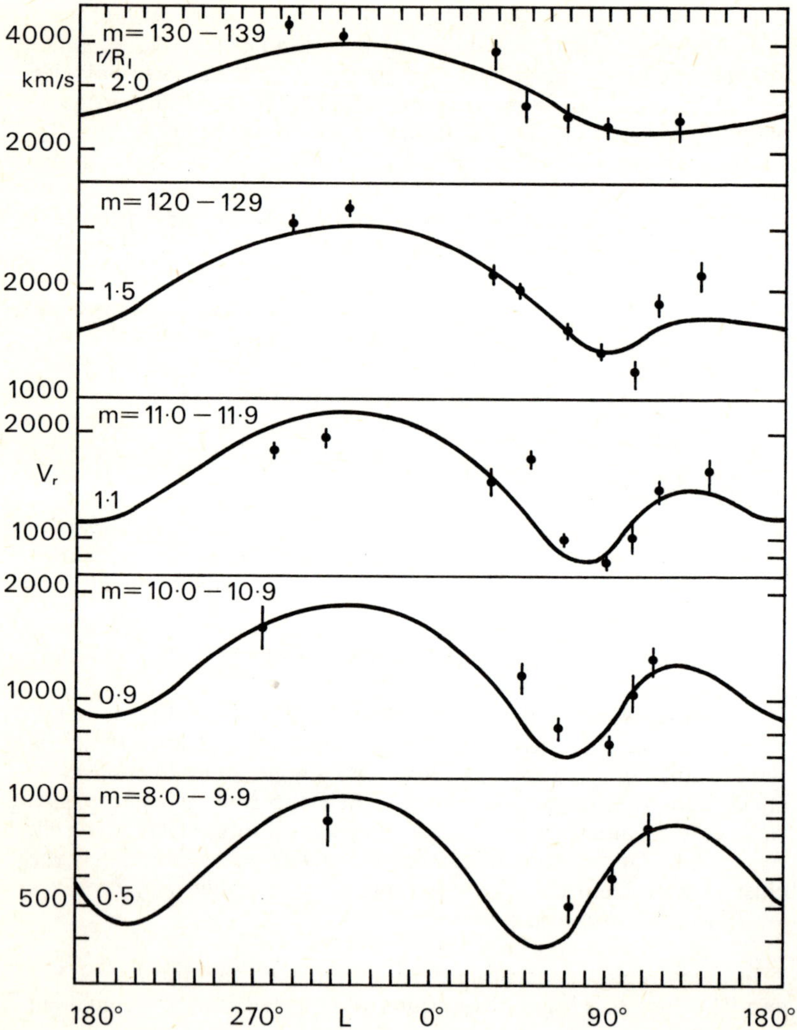
m= 130 − 139
r/R_I
4000
km/s
2·0
2000
m= 120 − 129
2000
1·5
1000
m= 11·0 − 11·9
2000
V_r
1·1
1000
m= 10·0 − 10·9
2000
1000
0·9
m=8·0 − 9·9
1000
500
0·5
180°
270°
L
0°
90°
180°

58 If the distribution shown in figure 57 is interpreted as marking a vast system of galaxies in rotation (that is, a supergalaxy), it is possible to give the longitude of an object belonging to the supergalaxy. In a rotating system, the velocity observed from any point in the system depends on the longitude and on the distances of the objects observed. The galaxies belonging to the supergalaxy have been separated according to magnitude intervals (corresponding, broadly speaking, to their distances) and their radial velocities have been given as a function of longitude. For the five distance-intervals considered, the theoretical curves have been adjusted according to the velocities observed (after de Vaucouleurs). At first sight, the result seems fairly conclusive, and the idea of a supergalaxy seems to agree with the aspect shown in figure 57. However, a more detailed statistical analysis, taking into account both the different distances of objects of the same apparent, but different absolute, magnitude, and also errors in the velocity measures, casts doubts on de Vaucouleurs' result. More research is needed into the whole question of a possible supergalaxy.

that the dense clusters are rich in intergalactic dust, which would absorb the light coming from more remote systems; this was particularly so for the Virgo and Coma clusters, and the cloud of galaxies in Ursa Major. A map of the clusters in the Ursa Major region, in which the clusters are arranged in order of distance, shows that remote clusters seem to avoid the areas in which there are closer systems. The irregular distribution is also brought out in the map of the Corona Borealis region, in which each galaxy is represented by a point.

The tendency to grouping

Studies of double galaxies give clear indication of the tendency of galaxies to form groups, and of the tendency for sub-groups to form inside clusters.

If the galaxies were distributed randomly throughout space, there would be cases in which two galaxies would appear close together simply because they happened to lie in much the same direction as seen from Earth; this does actually happen, and such pairs are called optical pairs. If galaxies between magnitudes 12·1 and 13 are taken from the Shapley-Ames catalogue, the numbers

Table 15 Numbers of physical and optical pairs among galaxies between magnitudes 12·1 and 13 (Shapley-Ames catalogue)

Separation (minutes of arc)	**Number of pairs**	**Optical pairs**	**Physical pairs**
0–4	35	1·4	33·6
5–9	35	5·0	30·0
10–14	37	8·6	28·4
15–19	37	12·1	25·9
20–24	30	15·7	14·3
25–29	31	19·2	11·8
30–34	27	22·8	4·2
35–39	33	26·4	6·6
40–44	37	29·9	7·1
45–49	31	33·5	(—2·5)
50–54	37	37·5	(—0.5)

of optical pairs can be listed, and the numbers of the physical pairs can then be obtained by simple subtraction. The results are given in table 15. Of course, the negative numbers have no significance, and their presence in table 15 is due merely to the small number of objects in each group. From the table, it is clear that the reality of physical pairs cannot be doubted, and when the separation is small the vast majority of the pairs are made up of galaxies which are physically associated. When the clusters are excluded, it is found that the maximum separation for a physical

Table 16 Frequency of systems in N (near) galaxies

Number of galaxies	Frequency
1	0·47
2	24
3	15
4	7
5	4
6	2
7	1

Table 17 Numbers of clusters of given component population

Number of component galaxies	Number of clusters
50–79	1,224
80–129	383
130–199	68
200–299	6
Over 300	1

pair of galaxies is about 225,000 parsecs.

The method can also be used to find out which of the multiple galaxies are optical and which are physical. Disregarding the clusters, the results are as in table 16. It seems, therefore, that on an average each galaxy has 1·1 companions, but this figure grows according to the increasing density of distribution. In the Virgo cluster, each galaxy is accompanied by an average of three physically associated galaxies, and the mean distance between the components of multiple galaxies is 85,000 parsecs.

59 A map of clusters of galaxies in a certain region of the sky (after Zwicky). Note the large clusters (11h 08m – 29°, 11h 13m – 29°) which are relatively close and which conceal more distant clusters. Note that these particular areas seem to include fewer clusters than other regions.

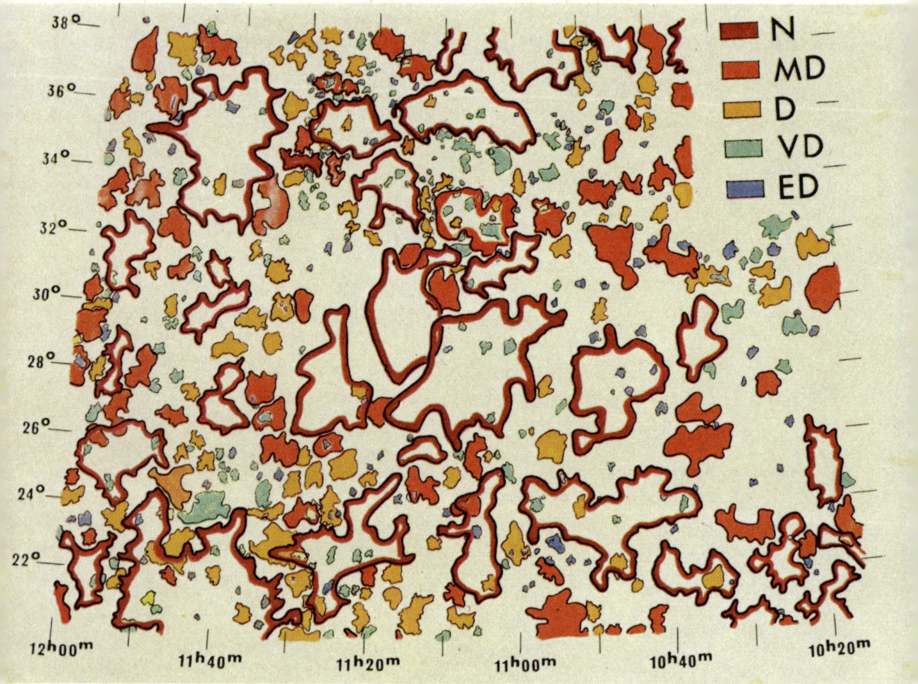

Large clusters

Various precautions must be taken in defining what may be called large clusters, because any cluster is bound to be seen in the same direction as other clusters both nearer and more remote, and it is not always easy to sort the various clusters out.

The number of galaxies in a cluster can be defined, in a purely conventional manner, by giving the number of galaxies brighter than a stated magnitude. Hubble selected 1,682 clusters from his catalogue, and worked out the frequencies of clusters in terms of the number of component galaxies included (table 17). In this list, all the clusters are within a distance of 600,000,000 parsecs, so that their velocities of recession are below 60,000 km/sec. The centres of these clusters do not seem to be distributed at random; in other words, there are apparent correlations between the positions of the centres of the different clusters in the catalogue.

Over the total of the regions studied, different methods can be used to compare the observed distribution with a completely random distribution. Abell has concluded that all correlation between galaxies ceases over distances greater than 40,000 parsecs. However, Zwicky, using other procedures, concludes that true clusters of galaxies do not exist over distances greater than 20,000 parsecs. Certainly the statistical analysis is difficult to make exact, and to give an idea of what is involved it will be helpful to digress for a moment into the theory of probability.

Statistical methods

Let us suppose that a certain number of balls is to be distributed at random in a certain number of boxes. All we are told is that the balls will not be influenced in any way, so that at the end of the experiment there may be boxes without a ball inside, or with one, two, three or more. The mean frequency is called Poisson's frequency. For example, with 2,000 balls and 1,000 boxes, the calculated mean numbers are:

BALLS		BOXES
0	in	138
1		271
2		271
3		180
4		90
5		36
6		12
7 and over		4

Obviously, if the balls were not independent of each other, the frequencies for zero, one, two balls and so on would be completely different. If, for example, balls were acceptable only in pairs, no box could contain an odd number of balls, and the number of empty boxes would be greater. For 1,000 pairs of balls and 1,000 boxes, the numbers would be calculated as follows:

BALLS		BOXES
0	in	368
1		0
2		368
3		0
4		184
5		0
6		61

It is therefore easy enough to decide whether a given distribution of balls in the boxes is random or not.

This sort of method of analysis has been applied to clusters of galaxies. The sky is divided into parts which are equal in area, and the number of clusters in each area is counted, to see whether the number of areas containing 0, 1, 2 . . . clusters of galaxies is the number which would be expected with random distribution. The results show conclusively that the clusters of galaxies are not independent of each other.

The selected areas must not be too large or too small, as in such cases the grouping of galaxies is masked. With very small areas, each area would be expected to contain either one galaxy or else none at all, while with very large areas the total numbers of galaxies would also be large and unwieldy. It is essential to take a happy mean for the size of selected areas. Abell has done this, and has found that the maximum 'distance for correlation' between clusters is 40 million parsecs.

Mean density

The mean density of matter in the universe may be estimated from studies of the mean number of galaxies per unit volume, and the mass/luminosity relationship for galaxies. The number of galaxies in any particular volume of space is highly significant, but unfortunately it is not yet known with any great accuracy. Not enough is known about the mean number of galaxies per cluster, the mean number of clusters per unit volume, the relative numbers of clusters of various types, and the mass/luminosity relationship for galaxies of different types.

One way to tackle the problem is to use Hubble's relation, which gives the number of galaxies per interval of magnitude. Extrapolation to fainter magnitudes, as observed in the Virgo cluster, leads to an estimate for the luminosity of galaxies per unit volume. Adopting the distance values given by Hubble's law, with the constant H = 100 kilometres per second per million parsecs, we find:

$$1 = 5{\cdot}1 \times 10^{-10}\ \mathrm{L}\odot\ \mathrm{pc}^{-3}$$

Adopting a mean value for the mass/luminosity relation equal to 21, as given by the Dutch astronomer Oort in 1958, the mean density is:

$$\bar{p} = 7{\cdot}3\ .\ 10^{-31}\ \mathrm{g/cm}^{-3}$$

A second method is to accept the results of statistical analysis of clusters of galaxies, giving about 10^{-19} galaxies per parsec. Also adopting Hubble's values for the distribution of galaxies per magnitude, and a value of 20 for the mass/luminosity relationship of the galaxies, it is possible to work out the luminosity per cubic parsec, keeping the same value for Hubble's constant:

$$1 = 3{\cdot}1\ .\ 10^{-10}\ \mathrm{L}\odot\ \mathrm{pc}^{-3}$$

With a mass/luminosity relationship of 20, the mean density becomes:

$$\bar{p} = 4{\cdot}6 \;.\; 10^{-31} \text{ g/cm}^{-3}$$

This result is naturally dependent upon the scale of distances, and this scale in turn depends on the value of Hubble's constant. If the value of this constant is changed, then the scale of absolute magnitude must also change, which affects both the number of galaxies per unit volume and the mass/luminosity relationship. If Hubble's constant is multiplied by a (for example, from 100 km/sec per million parsecs to 100a km/sec per million parsecs), the density is multiplied by a^2.

The two values for the density given here are of the same order, even though they depend upon different methods of investigation. However, a value of the order of a fraction of 10^{-30} grams per cubic centimetre depends entirely upon the value adopted for the mass/luminosity relationship of the galaxies. As has been noted, very high values for the mass/luminosity relation are found when the masses of galaxies in clusters are worked out on the assumption that the system is in a state of equilibrium. These high values have been questioned on other grounds, but we cannot quite exclude the possibility that the mean density of the universe is 25 times greater than the figure quoted above. This would give a result of 10^{-29} grams/cm^{-3}.

Olbers' paradox

The brightness of a luminous source falls off as the inverse square of the distance from the source. According to nineteenth-century views, this property could be extrapolated to infinity. It was also thought that the stars were uniformly distributed in space throughout the universe, and these two ideas led to a famous paradox which the German amateur astronomer H. Olbers published in 1826.

Suppose that there are N stars per unit volume, each of the same luminosity L. Divide space into spherical layers, of successive thicknesses δr_1, δr_2 and so on. In one layer there must be $4\pi r^2 \delta r N$

stars. At the centre of the sphere, each star will seem to shine with a brilliance which is inversely proportional to the square of the star's distance from the centre; the mean value will be $\frac{1}{2}(L/4\pi r^2)$. Stars in the spherical half-layer of thickness δr will produce a total brightness of $(L/4)\ N\delta r$. If layer is added to layer, right out to infinity, then the total illumination from the central observation point must also be infinite. This is obviously not the case, and so the next step must be to compare the calculated result with the actual brightness of the night sky. The total illumination of the night sky is equivalent to 10,000,000 stars of the 10th magnitude, and the combined illumination of all the galaxies out to a distance of 10,000 million light-years should be equivalent to 12,000 stars of the 10th magnitude. If the geometrical properties of space are ignored, then in a non-Euclidean universe the contribution of all the galaxies out to 10 million million light-years would give a luminosity equal to that of the whole night sky.

The feeble luminosity of matter in the universe can explain the small contribution made by the galaxies to the brightness of the night sky, but from a logical viewpoint Olbers' two hypotheses – that light decreases as the inverse square of the distance, and that luminous sources are uniformly distributed in the universe – lead to an absurdity, because they indicate that the sky should be infinitely bright.

Olbers resolved the paradox by introducing the idea that the universe contains absorbing matter, and even very slight absorption would be sufficient to hide remote objects completely. However, according to modern views this is no solution at all, because in an infinite universe the absorbing material would have infinite thickness; under conditions of equilibrium, the material would emit as much radiation as it absorbed, and the end product would still be a sky of infinite brightness.

Olbers' paradox is founded on a few definite assumptions, as follows:

1 The mean density-distribution of the stars is constant throughout space.

2 The mean density-distribution of the stars is constant in time.
3 The mean luminosity of the stars is the same throughout space.
4 The mean luminosity of the stars is invariable in time.
5 There are no systematic or co-ordinated movements of the stars.
6 The laws of geometry as known on Earth are valid throughout the universe; in particular, the apparent brightness of a source of light decreases as the inverse square of the distance of the source.
7 Physical laws as known on Earth are valid everywhere, even for the greatest possible scales.

In reality, there can be no doubt that Assumption 4 is untenable, because the energy reserves of the stars are not infinite. If we postulate a galaxy whose whole mass is composed of hydrogen, and whose mass/luminosity relationship is 20, the time needed for complete exhaustion of the hydrogen will be 200 thousand million years. This idea brings in several more points. It is reasonable to suppose that the stars in a galaxy can eject material which will be broken up and used in the formation of new stars, so that fresh generations of stars are born from nebulous material sent out from the old-generation stars. As has been noted, the newer material will differ from the older insofar as chemical composition is concerned.

Assumption 5 has not been verified. The red shifts in the spectra of galaxies are in themselves sufficient to resolve Olbers' paradox, because it can be shown that galaxies below magnitude 21 contribute up to one-half the total luminosity provided by all the galaxies – and the total due to all the galaxies amounts to only 0·3 per cent of the brightness of the night sky.

There is no good reason to think that Assumption 6 is valid; and in any case, general relativity shows that the laws of Euclidean geometry are not valid when the scale becomes very large. More about this will be said later.

Neither can Assumption 7 be verified, because the laws of motion are altered when the scale is sufficiently great. All in all, it is plain that even theories about the luminosity of the night sky can lead on to considerations of the universe taken as a whole.

Radio sources

Olbers' paradox turns up again with radio sources, with the added difficulty that it cannot be resolved by bringing in the red shift. For very high frequencies, perhaps even those of X-rays, the intensity of the radiation decreases slowly, proportionally to the frequency, to the power —0·6 to —0·8: I is proportional to ν^{-p} (using the previously-accepted values of p). If radio sources which are equal in power and are distributed throughout space to infinity are not to make an infinitely great contribution to cosmic radio noise, then the intensity must decrease more rapidly than the inverse of the frequency. Out to a distance of ten thousand million light-years, the contribution of all the radio sources should be 10^{-21} watts per cycle per second up to 400 megacycles, which would be ten times the observed contribution from all the discrete radio sources. If the red shift law were valid out to infinity, the contribution of all the discrete radio sources out to ten million million light-years would be 400 times as great as the flux which is actually observed.

These figures show up the contradictions very well, but there are other facts to be borne in mind. The red shift law cannot really be extrapolated so simply, and assumptions 6 and 7, concerning the validity of Euclidean geometry and terrestrial physical laws under all possible circumstances, must be summarily abandoned. Actually, observation does not indicate that the sources are distributed in the way that Euclidean geometry would have us expect.

The flux received from a source should be proportional to the inverse square of the distance, and the number of sources, above a certain limit, should be proportional to the cube of the distance, and also, therefore, to the $-3/2$ power of the flux:

$$N = \text{cte } F^{-3/2}$$

Such a distribution would give an infinite value for the total flux, as indicated by Olbers' paradox. Obviously, this does not happen. The observed distribution increases less rapidly when the

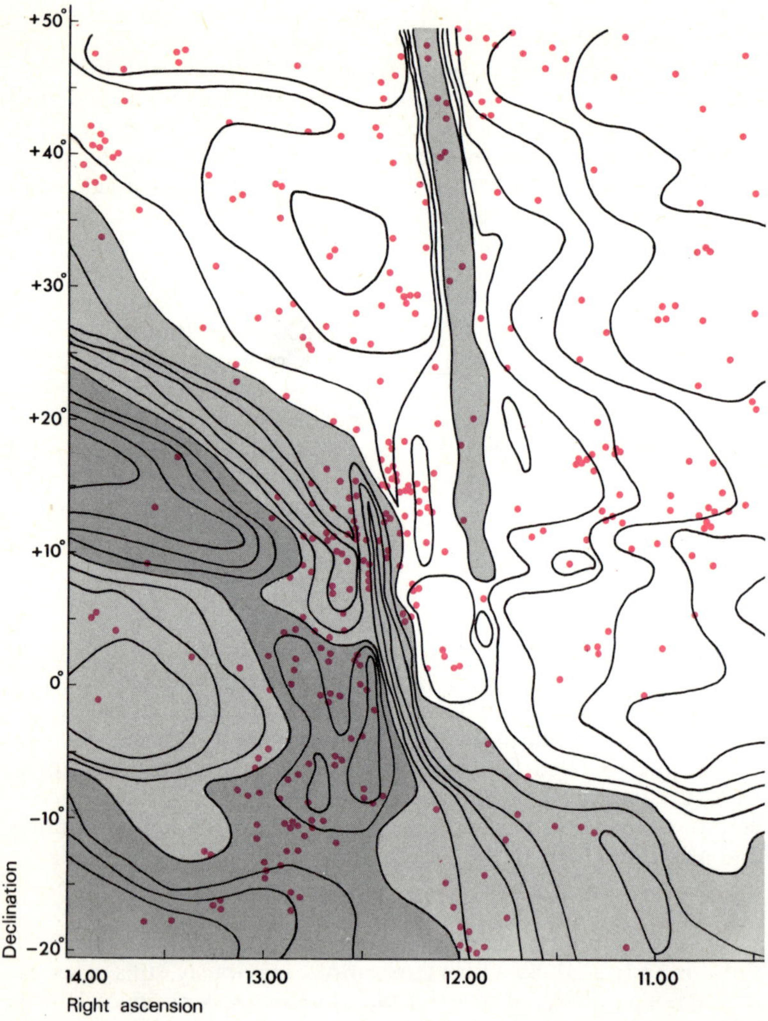
+50°
+40°
+30°
+20°
+10°
0°
-10°
-20°
Declination
14.00
13.00
12.00
11.00
Right ascension

60 Radio results at 180 cm (after Baldwin and Shakeshaft). The diagram shows the radio temperature curves for areas eight degrees square, the grey regions indicating more intense radiation. The galaxies in the Shapley-Ames catalogue are shown on the same diagram. The increase in flux in the area between 10h 30m and 13h is due to the radiation produced by the Galaxy. The intense area between 12h 30m and 14h, and 8° and 27°, is a region of galactic emission. The emission area at 12h, extending northward away from declination +10°, is due to the strong source Cassiopeia A. The less extensive area from 12h 30m, 12°, is due to the source Virgo A and the intense region at 12h 15m to 13h, 15° to +5°, may be due to a supergalactic effect. However the existence of a supergalaxy can be established only by other observations at other wavelengths. Shapley, who coined the term 'supergalaxy', considers that the only significance of this distribution is a perspective effect of a thousand close galaxies seen against a background of thousands of millions of more distant galaxies.

flux decreases, so that the total number of sources corresponds to the observed total flux, which is, of course, finite.

From the cosmological point of view, the main interest of the radio sources is that they can be detected out to enormous distances. More will be said about the observational data in the section of this book dealing with models of the universe.

Cosmic rays

In the first section, it was noted that cosmic rays cannot be permanent members of our Galaxy. The total flux of cosmic ray particles on the Earth's surface is of the order of 0·6 particles per square centimetre per second, corresponding to a density of 1 particle per 50,000 cubic metres. If the density of cosmic ray particles is taken to be constant throughout the universe, the corresponding density of energy works out to around 1 erg/cm^3.

It is well known that Einstein's formula $E = mc^2$ establishes a link between mass and energy. If the energy of cosmic radiation is known, its equivalent in terms of mass can be calculated, and is found to be approximately 10^{-33} grams per cubic centimetre, that is to say, about 1/500 of the density of matter as calculated from studies of galaxies.

It is important to find out how long the cosmic-ray particles travel through intergalactic space. Modern theories allow us to fix a maximum possible value for the density of the gaseous material in the universe, and it is known that to stop a particle with an energy of a million million electron volts would need encounters with 10^{32} atoms. If a cosmic-ray particle were to travel for a thousand million years without being halted, it would certainly encounter less than 100,000 particles per cubic centimetre. This enormous number can hardly be found anywhere in intergalactic space, and we must look for another solution.

If cosmic ray particles cannot be braked by molecules of gas, it is worth considering whether they might be slowed down by neutrinos or photons. Neutrinos are relatively sparse, with a density of only one per 400 cubic metres, but there are about 1,000 photons per cubic metre; the density of residual atoms cannot be greater than one atom per 40 cubic metres. Taking an effective section of about 10^{-26} square centimetres, it seems clear that the freedom of a cosmic ray particle is limited essentially by photons. The free path, or 'clearway', is of the order of one hundred thousand million light-years.

Cosmology

There is no point in giving further discussions about the properties of the universe without introducing cosmological problems, and, after all, cosmology is the science which deals with the universe considered as a whole.

Olbers' paradox, both in its optical and radio sense, means that we have to try to describe the universe in regions which are beyond our observable range; we must also bear the quasars closely in mind, because their speeds of recession are so staggeringly great. There is an immediate problem to be faced, because it is necessary to abandon Euclidean geometry and apply, to three-dimensional space, ideas which have been borrowed from the properties of two-dimensional surfaces. This is not an easy matter, and it is bound to cause a certain amount of mental confusion at first. Cosmology sets out to attack the problems of conditions at the limits of the universe if we regard the universe as finite, and the conditions at infinity if we take the contrary view. New physical concepts are needed, and these lead in turn to new theories and new ideas.

The idea of curvature

Consider a curve drawn on a plane surface. At each point, it is possible to define the tangent to the curve, and the normal to this tangent. Two normals close to each other intersect at a point which may be taken as the centre of curvature for that particular part of the curve. Working from this point as a centre, a circle can be drawn which just touches the curve and which, at the point of contact, is very like the curve itself. The distance from the centre of curvature to the tangent is the radius of the circle, and is called the *radius of curvature* of the curve at this particular point. The radius of curvature is, of course, constant at every point on a circle.

Next, consider a surface within a three-dimensional space. Creatures living on the surface would regard it as a two-dimensional space. An infinite number of curves may pass through any two points of a surface, but, generally speaking, one of these is bound to be the shortest possible curve; it is known as the *geodesic* of the surface. The geodesics of planes are straight lines, in keeping with the famous definition of a straight line as being the shortest distance between two points. Geodesics are simply extensions of this principle; for example, the geodesics of spherical surfaces are great circles. Only one great circle can be drawn through two given points of such a spherical surface, unless the two points are at opposite ends of a diameter, in which case, the number of great circles that can be drawn through the two points is infinite.

An infinite number of geodesics can pass through any one point. When considering tangents to geodesics, we are able to choose two geodesics whose tangents are at right angles to each other, and hence to define two geodesics which cut each other at right angles.

61 Equal-density contours in galaxies (after Shane *et al*). Note the contours which show density peaks, due to the presence of clusters of galaxies.

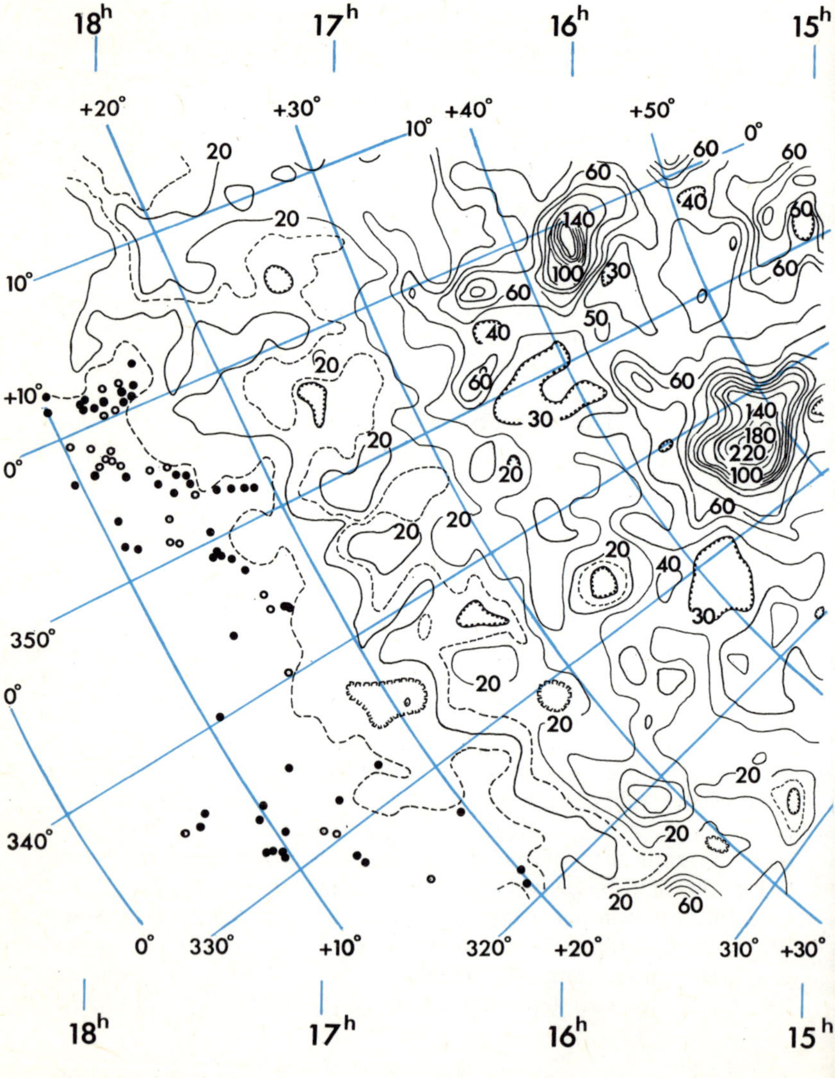

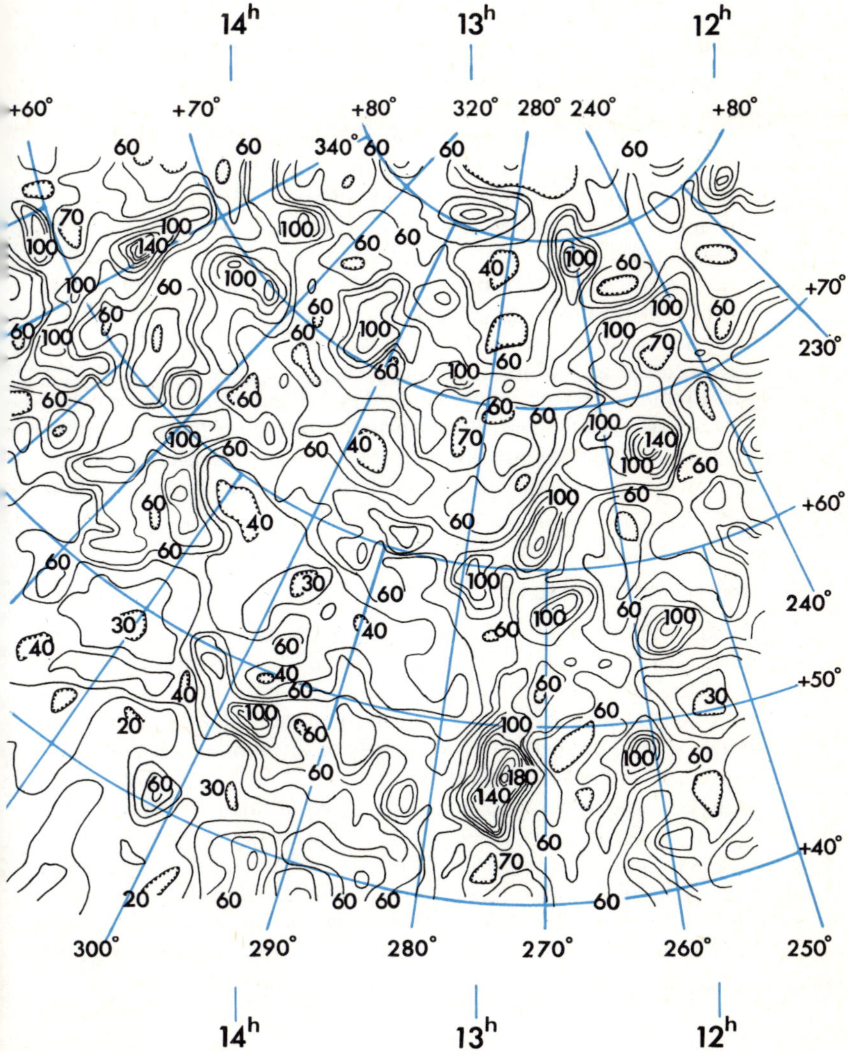
14h
13h
12h
+60°
+70°
+80°
320°
280°
240°
+80°
340°
+70°
230°
+60°
240°
+50°
+40°
300°
290°
280°
270°
260°
250°
14h
13h
12h

A normal plane can describe an arc of a curve on a surface. This curve will have a radius of curvature and a centre of curvature. If the plane rotates continuously around the normal, it will cut off different arcs of the curve, and the curvature will be different in each case. In general, the centre of curvature will occupy two extreme positions at right angles to the normal plane, and these two extreme positions define the two principal radii of curvature of the surface. Clearly, there will be a difference between surfaces such as spheres, for which both the two principal radii of curvature have the same sign, and hyperboloids, for which the principal radii of curvature have opposite signs.

The sphere is a special surface for which the two radii of curvature are equal at all points of the surface. It is a closed surface, and is also a surface of positive curvature. The hyperboloid, on the other hand, is a surface of negative curvature. Following the great nineteenth-century mathematician Gauss, we can define a scalar curvature, equal to the inverse of the product of the principal radii of curvature.

Geometry on a surface

The ordinary principles of plane geometry can be adapted for geometry on a surface. Replacing straight lines by geodesics means that triangles, quadrilaterals and equal lengths can be drawn, and the idea of a circle can be generalised by considering the figure obtained when equal segments of the geodesic are laid off, starting from one definite point.

Let us take the sphere, for example. If equal arcs of great circles are drawn, starting from a set point, the resulting points give us a circle. This circle is a *small circle* on the sphere, and it is at once clear that for this small circle, the ratio of the circumference to the radius is less than 2π. When the 'radius' is increased, the ratio becomes less. When the 'radius' amounts to one-quarter of a great circle, the 'circle' becomes the equator of the sphere, and the ratio of perimeter to radius is 4, again less than 2π. When the 'radius' is

62 Two planes AB, A′B′ passing through the normal PN cut two curves K, K′ on a surface. The centres of curvature are respectively C and C′.

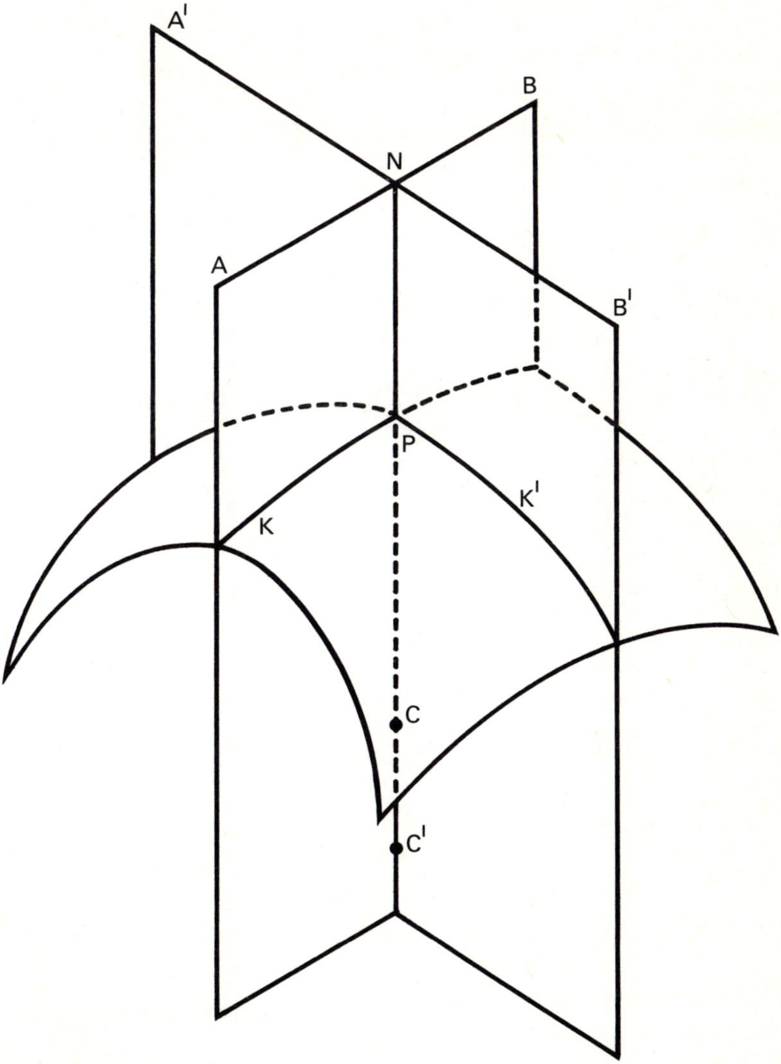

63 From point P there are arcs of equally great circles PA, PB, etc, or PA′, PB′, etc. The extremities of the arcs of the great circles describe a small circle C, C′, etc, for which the ratio of the circumference to the 'radius' PA is less than 2π.

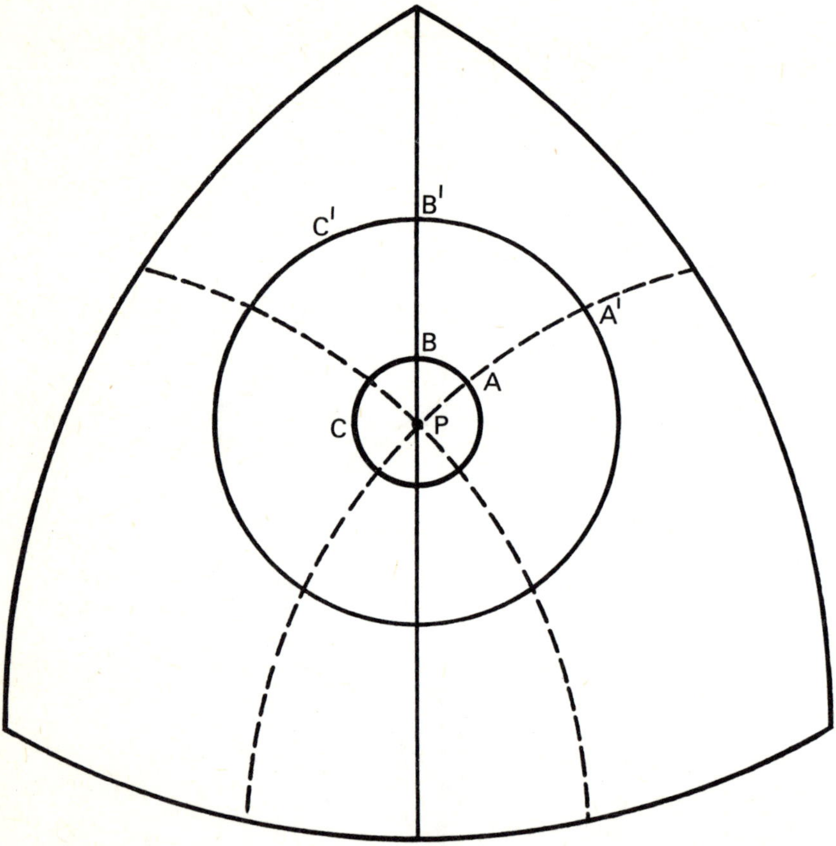

64 Tracing a 'rectangle' on a sphere. The 'rectangle' is made up of arcs of great circles, as described in the text, but the 'rectangle' is not closed because point E does not fall on top of point A.

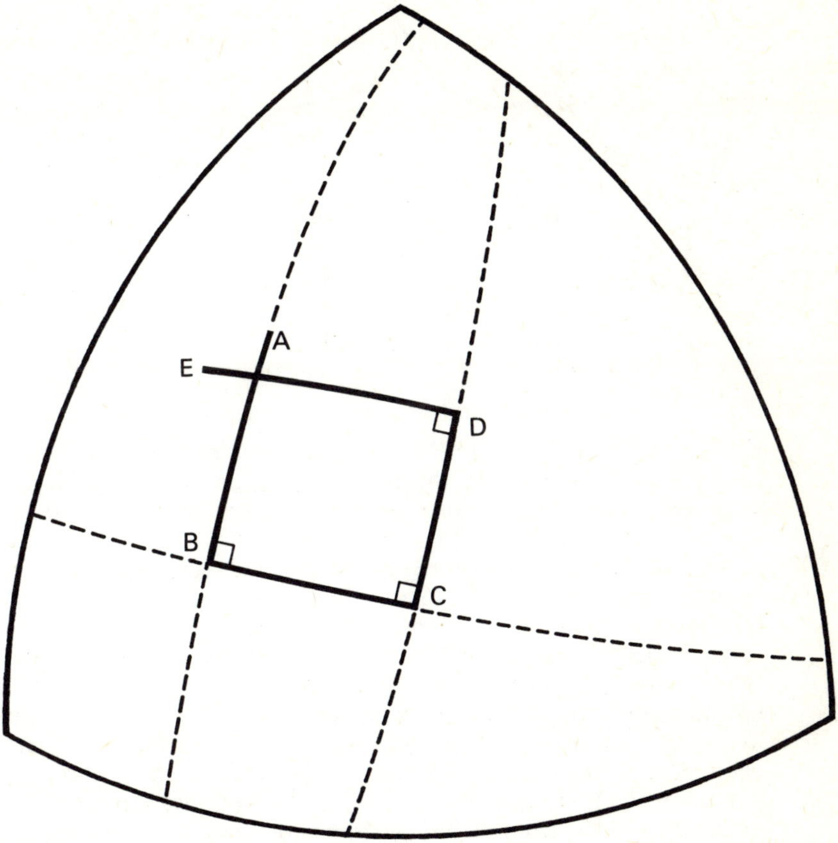

65 Tracing a 'rectangle' on a hyperbolic surface.

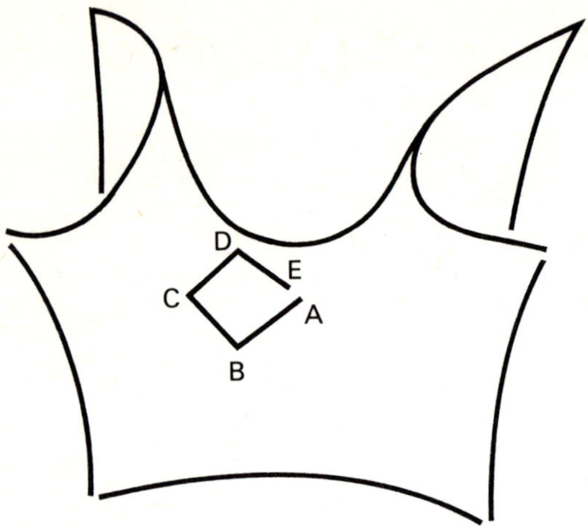

half a great circle, the 'circle' is reduced to the pole opposite to its centre – and the ratio of the perimeter to the 'radius' becomes zero. On a sphere, the ratio of the perimeter of a circle to its radius must always be less than 2π.

Another experiment with regard to the curvature of a sphere is to draw a 'rectangle' or a 'square' on it. Mark off a small arc AB on the great circle of a sphere, and then try to draw an arc BC equal in length to AB and at right angles to it; continue with similar arcs CD and DE – and it will be found that point E does not coincide with point A. In fact, it is impossible to construct a 'rectangle' made up of segments of geodesics which are perpendicular to each other. The distance between points E and A is due upon the curvature of the surface, and becomes greater when the lengths of the sides of the 'rectangle' or 'square' are increased.

On a hyperbolic surface (such as the relief of a mountain in the neighbourhood of a pass), the ratio between the perimeter and radius of a circle is always greater than 2π. On a sphere, the sum of the angles of a 'triangle' is always greater than two right angles, but on a hyperbolic surface the sum of the angles of a 'triangle' must always be less than two right angles.

In theory, it would be possible to work out the form of a hill by drawing geodesics and 'rectangles' in it, by analysing the ways in which the rectangles failed to close; it would be possible to estimate the curvature of the hill, whether the observer happened to be near the summit or in the pass below. Geometers living in a permanent fog, and unable to see the shape of their mountain, could nevertheless find out its exact form simply by drawing mutually perpendicular geodesics on the ground.

Geometry in space

In the neighbourhood of the Earth, the geometry of space is to all intents and purposes Euclidean. The great mathematician Gauss was able to show that even with a very large triangle, with sides about 100 kilometres long, the sum of the angles is equal to two right angles. And yet there is no *a priori* reason to believe that when immense distances are involved, space remains Euclidean. In other words, if an attempt were made to trace our 'rectangles' in space, the rectangles would not close; the sum of the angles of a triangle would not be equal to two right angles, and the ratio of circumference to radius would not be equal to 2π. The idea of there being a shortest distance between any two points must of course be retained, but because the 'rectangles' do not close it must be said that space is *curved*.

The essential aim of general relativity is to link the geometrical properties of space with the distribution of matter in space. In relativity, the shortest distance between any two points is the path which would be followed by a ray of light, so that light-rays in space always follow geodesics.

The three proofs of relativity

Relativity is a new theory of gravitation, which relates the properties of matter to the geometrical properties of space. In relativity, gravitation is propagated through space with a finite velocity: the

Table 18 Schwarzschild's singularity

Mass (Sun=1)	Radius of Schwarzschild singularity (km)	Mean density
1	3	$1{\cdot}8 \times 10^{16}$
100	300	$1{\cdot}8 \times 10^{12}$
10,000	30,000	$1{\cdot}8 \times 10^{8}$
1,000,000	4·3 solar radii	$1{\cdot}8 \times 10^{4}$
100,000,000	430 solar radii	1·8

velocity of light. The theory also sides with Descartes against Newton, in that it affirms that there is no action at a distance. Three observational proofs of the truth of relativity theory have so far been found, and these are so important that they must be described in some detail.

1 Close to the Sun, space no longer obeys the Euclidean laws, and geodesics are not straight lines. As soon as Einstein's general theory of relativity was published, scientists started casting around for observational proofs, and the obvious one concerned the apparent positions of stars very near the Sun in the sky. If it were possible to measure the position of a star first when well away from the Sun, and then when almost behind the Sun, the difference ought to show up; the light-rays would be deviated, because when passing near the Sun their paths would not be straight lines. The main difficulty is that the stars cannot normally be observed in broad daylight, but during a total solar eclipse the sky becomes dark, and stars close to the Sun can be photographed. This was done at the total eclipse of 1919. When the measured star-positions were compared with the positions on a photographic plate taken six

months later, when the Sun was out of the way, the expected differences were duly found, and this was a real triumph for relativity theory. When light-rays pass near the Sun, they are slightly curved. The maximum deviation is 1·7 seconds of arc.

2 If a photon is to be taken out of a gravitational field, a certain amount of work must be done, and this work comes from the energy of the photon, with a consequent increase in the photon's wavelength. In fact, the result is a gravitational shift of the photon toward the red end of the electromagnetic spectrum. The shift is slight; expressed in the same way as a Doppler red shift effect, its value at the surface of the Sun would amount to only 0·6 kilometres per second, which means that it can be measured, but with considerable difficulty. At the surface of a white dwarf star the value can go up to 20 kilometres per second, which can be measured with relative ease.

3 From the viewpoint of an observer situated at a great distance, the Sun's gravitation, in an Euclidean system, differs slightly from the field as laid down by Newton's law (that is to say, gravity falls off as the inverse square of the distance). The difference modifies the movements of the planets to some extent; the effect is most noticeable for Mercury, whose perihelion advances by 43 seconds of arc per century. The effect was discovered in the nineteenth century by the French astronomer U.J.J.le Verrier, though it was not explained until much more recent times.

Schwarzschild's singularity

The curvature of the paths of light rays in the neighbourhood of the Sun is very slight, but only because the energy of the Sun's gravitational field is extremely weak. If the mass of the Sun could be concentrated into a sphere of very small radius, the resulting curvature of space would become much greater; and if the mass of the Sun could be concentrated into a sphere of radius 3 kilometres, light could not leave the object, but would move on the surface in a circular path. Therefore, it seems that a star of solar mass but with a radius of only 3 kilometres must be invisible; no ray of light could leave it or reach it, simply because of its tremendous gravitation.

Table 18 gives the radius and mean density of such invisible spheres, as a function of their mass. The radius of the boundary inside which the material becomes invisible is called Schwarzchild's singularity in honour of K. Schwarzschild, the famous physicist and mathematician who discovered it.

Collapse or explosion ?

As so often happens nowadays, it is found that studies of quasars are significant in researches of this kind. The total energy of a quasar may be as much as 10^{62} or 10^{63} ergs, and it has seemed to cosmologists that the only means of liberating energy upon this grand scale is by the gravitational collapse of very massive bodies. Very roughly, the energy available would be of the order of the product of the mass by the square of the velocity of light, and it follows that the collapse of a body of about 100,000,000 times the mass of the Sun would release enough energy to explain the quasars. The question of collapse also turns up in investigations into the evolution of the stars. Under certain conditions, a star with a mass more than three times as great as the Sun's can apparently undergo indefinite contraction. When its radius has shrunk to the value of the Schwarzschild singularity, the star ought therefore to disappear.

When a mass lies within the Schwarzschild singularity, the properties of light which have been described earlier in this section result in several paradoxes involving collapse or explosion. A mass of material in the process of collapse under its own weight remains visible as long as light can escape from it, that is to say, while the radius remains greater than the value for Schwarzschild's singularity. The movement of the material and the movement of the light are in opposite directions, since the material is moving toward the centre and the light is travelling outward from the centre. An observer situated well away from the body will see that when the radius reaches the Schwarzschild value, the object will vanish. The paradox arises because for this outside observer, the body will take an infinite time to disappear.

If the sense of time is reversed, it seems that we appear to be dealing with an explosion, but the description as given above is no longer valid, because so far as the outside observer is concerned, both material and light will be moving in the same direction, that is, away from the centre. Therefore, our external observer would presumably see an exploding object emerge from the Schwarzschild singularity in a finite time. This corresponds more or less to an idea of Jordan, reconsidered recently by Neeman and by Novikov, in order to explain the eruptions of quasars.

Single galaxies

Ambartsumian has recently pointed out that all the single galaxies, radio galaxies and even our own Galaxy give indications of explosions, or at least expansion of material. In his view, there is little justification for the idea that these phenomena should be linked with the energy gained by the original contraction, even if gravitational energy were liberated during the course of this contraction. For this reason, it is unwise to disregard the theory that the quasi-stellar sources are caused by large masses of material emerging from the Schwarzschild singularity. Such masses could have a slowed-down expansion, as will be described later in this section.

An extra reason for trying to explain the quasars in this way is that when gravitational contraction is in progress, the binding energy of atomic nuclei is by no means negligible. On balance, it is not certain that contraction makes a major contribution to the amount of energy available, which is a significant point. Moreover, the paradox described above seems to show that so far as an external observer is concerned, an explosion out of the Schwarzschild singularity would necessarily occur in a finite time.

The curvature of space

The properties of 3-dimensional and 4-dimensional space are naturally more complex than those of 2-dimensional space (that is to say, surfaces). The properties of the curvature of a surface are determined by a single quantity, Gaussian curvature. The properties of 3-dimensional space are determined by six components, which may be reduced to three independent quantities; the properties of the curvature of 4-dimensional space are determined by no less than twenty components, which may be reduced to fourteen independent quantities.

However, for a medium which is homogeneous, that is to say, which has the same properties at every point, the number of constants needed to determine the properties of the curvature is reduced to 1, as in the case of surfaces. If this constant is positive, then 3-dimensional space has a constant positive curvature, and has properties analogous to those of a sphere; if the curvature is negative, then 3-dimensional space will have constant negative curvature, and its properties will be analogous to those of a hyperboloid.

Obviously, it is much more difficult to visualise the properties of curved space than to picture the properties of a sphere, but some idea of what is meant can be given by considering the properties of a sphere in curved space with a constant positive curvature. It is assumed that the sphere has been constructed by drawing geodesics of equal length in all directions. When the radius of the sphere is small, the ratio of the surface to the square of the radius is 4π, and the surface increases with the radius. When the radius reaches a value of πR, where R is the radius of curvature of the space, the surface of the sphere reaches its maximum, and the ratio becomes $(16/\pi) = 5{\cdot}1$. As the radius continues to grow still further, the surface of the sphere starts to shrink, and is finally reduced to a point at the opposite pole at a distance πR, the greatest possible distance in the space under consideration. The sphere constructed in this fashion fills the entire space, of volume $2\pi^2R^3$. Space with constant positive curvature is finite, closed on itself, and of finite volume; but obviously it has no boundary.

With space of constant negative curvature, the ratio of the surface of the sphere to the square of the radius is equal to 4π for spheres of very small radius, but increases steadily with the growth of the radius of the sphere. The volume of the sphere increases indefinitely with the increase of the radius, and the volume of space with constant negative curvature is infinite.

Newtonian cosmology

In Newtonian cosmology the universe is Euclidean; space has no curvature, and the properties of light are not linked with the properties of matter, as they are in relativity theory. In other words, everything happens as though the velocity of light were infinite; there are no limitations on velocities.

The Euclidean universe – homogeneous, filled with material of constant density, and in equilibrium – is not a possibility. To explain why this is so, let us consider a homogeneous sphere of radius r. The gravity at its surface will be $4/3\pi G\rho r$, and this will increase proportionately with the radius. For a sphere of infinite radius, the surface gravity is infinite. This is a physical impossibility. Two answers to the problem may be put forward; one introduces the idea of motion in the universe, while the other involves a modification of the laws of gravitation. Actually, these two solutions are not mutually exclusive.

An isotropic, homogeneous expansion will remove the difficulties facing Newtonian cosmology. At every point, the acceleration due to the expansion will exactly compensate the force due to weight. In the simplest case, the distance between any two points in the universe will increase according to the 2/3 power of the time; in this case, the date for the start of the expansion is linked with the density of the material by the relationship $t = (6\pi G\rho \text{ actual})^{-\frac{1}{2}}$, and with ρ actual $= 5 \times 10^{-31}$ g/cm^{-3}; the time t works out at 4×10^{10} years. The corresponding velocity of expansion is 16 kilometres per second per megaparsec, or about six times slower than the observed velocity of expansion. In other words the energy-density of the expansion of this Newtonian universe is about 40 times greater than the gravitational energy-density.

The law of gravitation can be modified so as to become a law of cosmical gravitation. This involves the introduction of another force, cosmical repulsion, which is the opposite of gravitation, and increases according to increasing distance. By suitable selection, a term of cosmical repulsion can be introduced so that the gravitational attraction will be exactly balanced, and there is a state of equilibrium – but this equilibrium is unstable, and the least perturbation will start the universe in a career of evolution. Increase in density means that there will be a catastrophic contraction, while a decrease in density will result in unlimited expansion. The reason for this is that when density is increased, the force of gravity overcomes cosmical repulsion; when the density drops, cosmical repulsion becomes more powerful than gravitation.

The law of gravitation can also be modified by the introduction of a gravity-screening term, and this can lead to a static solution. This solution has the peculiarity that it is stable with regard to every large-scale deformation, as though the gravity-screen masks the discrepancy between attraction and repulsion; in other words, if the system is originally stable there can be no subsequent general expansion or contraction, though there may be small local variations. An obvious difficulty facing the static cosmological model is its failure to explain the red shift. Instead of supposing that the universe is expanding, the theory must introduce a new idea, according to which a photon 'ages' and increases its wavelength. This would certainly account for the red shift, but so far as we can tell there is no justification for introducing a new physical law for which there is not the slightest evidence. It seems, therefore, that all static models of the universe must be rejected.

However, two special features of the gravity-screen model are worth noting. As has been pointed out by Zwicky, the model explains the almost complete absence of clusters of galaxies, and this determines the distance out to which the effects of the screen make themselves felt – of the order of 20,000,000 light-years. On the other hand, if gravitation is explained by introducing another elementary particle (the *graviton*), the law of attraction follows the Newtonian law exactly, provided that the graviton has no mass; if the graviton is assumed to have mass, then the effect of gravity would fall off, with increasing distance, more rapidly than in Newton's law. In the graviton theory, the effects of the screen seem to be linked directly with the mass of the graviton. It has been calculated that if the screen has a range of the order of 20,000,000 light-years, the graviton would have a mass of 10^{-62} grams.

It is true that the idea of photon ageing and the graviton hypothesis seem somewhat far-fetched, but they cannot be rejected out of hand. Moreover, there are many even stranger cosmological theories which have been put forward from time to time.

Einstein's cosmology

In relativistic cosmology, the geometrical properties of space are directly related to the distribution of matter. One of Einstein's essential ideas was that gravitation is not merely influenced by the distribution of matter, but is determined by it. On this view, it seems inadmissible that the solutions of the relativistic equations can be determined by conditions of limits; and the best way of getting rid of the embarrassment of limit conditions is to remove the limits themselves. As the equations gave a static solution only for an empty universe, Einstein, in 1916, introduced the cosmical repulsion term into his equations, and this led him on to a spherical static universe containing matter at a finite density. This universe was closed upon itself, finite but unbounded, and seemed very satisfactory, because there was no longer any obvious need to be bothered about limits. However, it was achieved only by the introduction of cosmical repulsion in a form which was admissible from a relativistic point of view but unacceptable from the logical viewpoint, because it was justified by no physical considerations. The cosmical repulsion term had been introduced only so that a static solution of the equations could be found, and Einstein later rejected the term.

Note that the radius of curvature of the spherical, static Einstein universe would be about 48,000 million light-years, when the density is taken as 5×10^{-31} grams per cubic centimetre.

Friedmann's cosmology

Friedmann was the first to give complete equations for a homogeneous and isotropic universe, and to discuss the various different possible solutions. His work was published in two mathematical memoirs, the first in 1922 and the second in 1924. He showed that Einstein's closed spherical universe must be the only possible static universe, because the equations did not allow a static solution for a finite universe containing matter at finite density. Above all, Friedmann showed that the existence of solutions depended upon time. Before the red shifts in the spectra of galaxies had been observed, they were to all intents and purposes predicted by Friedmann's solutions, according to which the radius of the curvature of the universe increases in the course of time.

In particular, Friedmann showed that Einstein's equations, without the mysterious cosmical repulsion term, could give many solutions, but all of these were non-static; in fact, in relativistic cosmology, as in Newtonian cosmology, there can be no such thing as a static-universe model, unless cosmical repulsion is brought back. Models in which the universe is assumed to be expanding seem to agree much more closely with the observed facts.

The next stage is to look for suitable quantities which may help in drawing up a model of the universe. Several quantities are accessible to observation: the red shift as a function of distance, the number of objects of given kinds which exist as a function of distance, and the mean density of matter in space. Only the first and last of these quantities are known with any accuracy. Counts are unreliable, and diameter measures are of little use unless they apply to identical objects. Therefore, the parameters of models of the universe are by no means easy to obtain from observation alone.

Consider, for example, a model of the universe in which there is no cosmical repulsion. If the model is an open one, the relation between Hubble's constant, the density, and the radius of curvature is:

$$\frac{C^2}{R^2} = H^2 - \frac{8\pi G\rho}{3}$$

If the model is a closed one, the relation becomes:

$$\frac{C^2}{R^2} = \frac{8\pi G\rho}{3} - H^2$$

Table 19 gives the values of the radius of curvature as a function of the density. H is taken as being 100 km/sec per megaparsec. The very small value found for the density strongly suggests that the universe is open and infinite. Its radius of curvature (negative curvature) is approximately equal to the product of the velocity of light by the reciprocal of Hubble's constant, and this seems to be due to the great predominance of the density of energy of the expansion over the density of the gravitational energy.

Cosmical acceleration

There is no *a priori* reason why Hubble's constant, which defines the velocity of expansion of the universe, should be independent of time. A variation in Hubble's constant in the course of time constitutes what is known as cosmical acceleration. If this acceleration is positive, the rate of expansion will grow steadily as time passes; if the acceleration is negative, then the rate of expansion will slow down. It is easy to measure cosmical acceleration by means of a straightforward magnitude without dimension. To show how this

Table 19

Mean density (g/cm^{-3})	Radius of curvature (thousand million)	Nature of model
$6{\cdot}66 \times 10^{-31}$	10·3	Open space (negative curvature)
2×10^{-30}	12·8	Open space (negative curvature)
6×10^{-30}	17·9	Open space (negative curvature)
$1{\cdot}8 \times 10^{-29}$	Infinite	No curvature (Euclidean space)
$5{\cdot}4 \times 10^{-29}$	9·6	Closed space (positive curvature)
$1{\cdot}6 \times 10^{-28}$	3·4	Closed space (positive curvature)

is possible, it will be best to consider a few examples of simple motion:

1 *Uniform motion.* Uniform motion is represented by the following relationships:

$$\begin{array}{lrcl} \text{distance covered} & z & = & vt \\ \text{velocity} & \dot{z} & = & v \\ \text{acceleration} & \ddot{z} & = & 0 \end{array}$$

Writing $H = \dot{z}/z = 1/t$, the acceleration factor is calculated by:

$$q_0 = -\frac{\ddot{z}}{zH^2} = 0$$

For uniform motion, the acceleration factor is zero.

2 *Uniformly accelerated motion.* A projectile which is rising vertically in the Earth's gravitational field has a uniformly retarded motion, while when it falls back to Earth its motion is uniformly accelerated. If the starting point in time is taken as the moment when the projectile reaches its highest point, and the downward direction is taken as positive, we have:

distance covered:	$z = \frac{1}{2}gt^2$
velocity	$\dot{z} = v = +gt$
acceleration	$\ddot{z} = +g$

Velocity can be replaced by the quantity H: $H = z/z = 2/t$. And for the quantity $q_0 = -(\ddot{z}/zH^2)$, we have: $q_0 = -\frac{1}{2}$. So for uniformly accelerated motion, the acceleration factor is $-\frac{1}{2}$.

3 *Motion of a projectile receding from the Sun.* In this case, if the velocity is zero when the projectile reaches infinity, the following relations are valid:

distance covered	$z = kt^{2/3}$
velocity	$\dot{z} = v = 2/3\ kt^{-1/3}$
acceleration	$\ddot{z} = 2/9\ kt^{-4/3}$

Taking the quantity $H = \dot{z}/z = 2/3\ .\ 1/t$, the acceleration factor is:

$$q_0 = -\frac{\ddot{z}}{zH^2} = \frac{1}{2}$$

4 *More complex motions.* In these cases, the acceleration factor depends on the time; it measures the acceleration of the motion from a given instant. For example, consider the motion of a projectile. If the origin is not taken as the instant when the projectile reaches the summit of its trajectory, we find at the initial instant:

$$q_0 = -gz_0/v^2$$

where q_0 measures the acceleration. If the motion is not uniformly retarded, and the negative z is ignored, it is seen that $q_0 > 0$. But

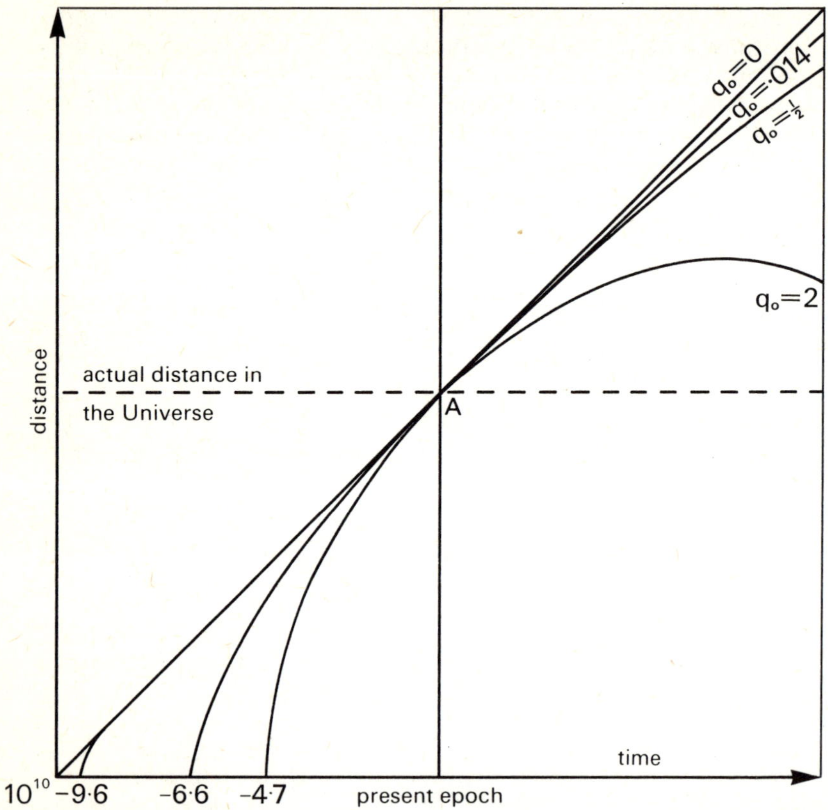
q₀=0
q₀=·014
q₀=½
q₀=2
distance
actual distance in
the Universe
A
time
10¹⁰ −9·6
−6·6
−4·7
present epoch

66 Variations in distances between galaxies in different models of the expanding universe. The variation of the distance *d* between two galaxies G_1G_2 (ordinates) is shown as a function of time (abscissa). The same point *A* represents the distance *d* and the actual instant of the cosmic time for all the models considered. The different curves correspond to different values of the parameter of acceleration. All the models considered here (no cosmological constant) have a negative or nil acceleration; in other words, the expansion is slow. The properties of the different models represented on the graph can be summed up as:

	Expansion	**Density**	**Model of the universe**
0	regular	$=0$	hyperbolic
0·014	slowed down	$=5{\cdot}10^{-31}$ (observed)	hyperbolic
1/2	slowed down	$=1{\cdot}8 \times 10^{-29}$	Euclidean
2	slowed down	$=7{\cdot}2 \times 10^{-29}$	spherical

Note that the spherical closed models with no cosmological constant are models of the oscillating universe, while the hyperbolic models are in continual expansion.

when the starting-point for both distance and time is taken as the top of the trajectory, we find, as before: $q_0 = -\frac{1}{2}$.

In cosmology, the variation with time of the radius of curvature R_0 has to be taken into account. Hubble's constant is $H = \dot{R}_0/R_0$, where $\dot{R}_0$ is the rate of change of the radius of curvature. Similarly, the acceleration $\ddot{R}_0$ is the rate of change of velocity. Expressing acceleration by the quantity q_0, then $q_0 = -\ddot{R}_0/\dot{R}_0\ H_0^2$. The quantity q_0 is positive if the motion is retarded, and negative if the motion is accelerated.

For a universe in which the cosmological constant is zero, the following relation can be established between the density of matter in space and the cosmical acceleration:

$$\rho_0 = \frac{3H^2{}_0q_0}{4\ MG} = 3{\cdot}67 \times 10^{-29}\ \mathrm{g/cm^{-3}}q_0$$

If the density ρ_0 which figures in this relation is really the same as that which figures in the expression of the curvature of the

67 The radial velocity magnitude ratio, according to Sandage. The dots represent observed clusters of galaxies; the curves represent models of the universe for different values of the parameter of acceleration.

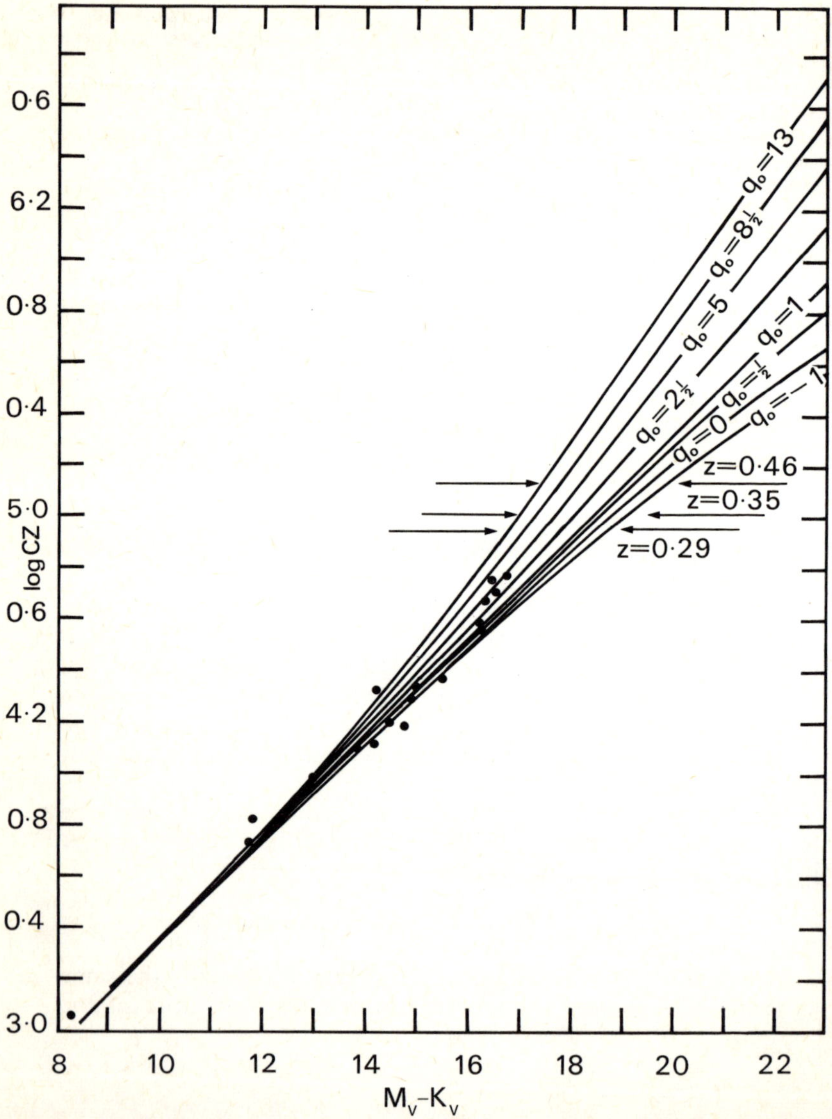

universe, it follows that:

$$\frac{k_c^2}{R_0^2} = H_0^2(2q_0 - 1)$$

in which k equals 1, 0 or −1 according to whether the universe is closed and Euclidian, or hyperbolic and open.

The diagram given here gives the shift as a function of magnitude, the theoretical relations calculated for different values of the acceleration term q_0, and the points observed. Data due to Humason, Mayall and Sandage seem to give something of the order of $q_0 = 2{\cdot}5$, but with great uncertainty. Baum's results are compatible with a value for q_0 between $\frac{1}{2}$ and 3/2. The lowest possible value for q_0 is zero, corresponding to a hyperbolic universe ($k = -1$). However, the values determined for ρ_0 seem to correspond to a density 70 times greater than the observed value. This could be dealt with by choosing a large, negative value for the cosmological constant, but unfortunately there is nothing to justify a choice of this kind. The value of q_0 calculated from the density-value is of the order of 0·014, which leads to an open hyperbolic model but which does not agree with the value found from observation.

Observations of quasars or very blue galaxies ought to clear up the uncertainty about cosmical acceleration, because these objects are luminous enough to be seen across immense distances. Certainly the spectra of the quasi-stellar sources are very different from those of other galaxies. Actually, the spectra of quasars seem to give a law linking power and frequency. For a law of this kind, it is possible to calculate the magnitude correction due to the red shift, and find a magnitude that has very little dependence on the shift. It is then an easy matter to work out what magnitude the quasar would have if there were no red shift. By taking the number of quasars as a function of their magnitude, there seems a good chance of obtaining a relation from which cosmical acceleration can be calculated. Clearly, everything depends upon our being able to observe quasars out to immense distances.

68 The 193-cm Cassegrain telescope at the observatory of Haute-Provence, France.

In America, Sandage undertook an investigation of this kind. He used a catalogue containing 8,746 very blue objects, and drew a curve giving the number of objects as a function of magnitude. He found that down to the 15th magnitude the number of objects increased very slowly, but below magnitude 15 the increase was much more rapid. Sandage concluded from this that the objects down to magnitude 15 belonged to our Galaxy, while the fainter objects were quasars. There were enough of the quasars to make possible a determination of the cosmical acceleration. It was very intriguing to see that the figure for the numbers of very faint, very blue objects seemed to be compatible with a value for the constant of acceleration $q_0 = 0$, though it did not absolutely exclude the value for $q_0 = 1$.

Unfortunately, recent efforts to verify the nature of these very blue objects, using a spectrograph, seem to show that only ten per cent of them are quasars. Sandage's results can therefore be questioned, but in principle his method is sound enough.

By now well over one hundred quasars have been identified, and more will have been tracked down by the time that this book appears in print. Eventually, it will certainly be possible to make counts which will lead to a much better value for the cosmical acceleration term. Certainly we should soon know whether the universe is hyperbolic and open ($q_0 \simeq 0$) or spherical and closed ($q_0 > \frac{1}{2}$).

Cosmic time

One of the strongest points of the theory of special relativity is the idea that time is relative. Consider an observer at rest, who sees a second observer passing by carrying a clock identical with his own. He will be able to notice that the second observer's clock is not keeping quite the same rate, because in special relativity there is no standard of absolute time applicable both to observers at rest and to those who are in motion. There is a contraction of the time-scale, which decreases with increased velocity. When the velocity becomes equal to the velocity of light, the contraction is 'complete';

the observer will go from one point to another at the velocity of light in zero time, though the observer at rest will record the passing of finite time.

This contraction of the time-scale was the basis of a concept described many years ago by Langevin, and known as Langevin's paradox. If an observer leaves Earth at a velocity so high that his time-scale is 100,000 times slower than that of a terrestrial observer, and if he comes back at the same velocity, two years will seem to pass by for the moving observer, but the terrestrial observer will find that 200,000 years have elapsed. More will be said about this paradox later in the present book.

The time-dilation effect has been experimentally verified by studies of the high-energy cosmic ray particles. Some of these particles have extremely short lifetimes, of the order of a thousand-millionth of a second. However, when they are moving at near the velocity of light, their lifetimes appear to be lengthened. In the experimental research, the life-spans of the particles are measured by the length of their tracks in the gelatine of a photographic plate; the longer the lifetime of the particle, the longer the track in the gelatine, and the overall aspect depends on the energy of the particle, which in turn is controlled by its velocity. It is found that the life-span is greater when the energy of the particle is greater, and the relationship between the two factors agrees excellently with the predictions of special relativity.

A more searching analysis of the nature of the time-dilation effect shows that, as Einstein said, it is due to the finite value of the velocity of light. In the final analysis, two physical systems are taken to communicate with each other by means of signals which travel at the velocity of light. As with other physical quantities, the rate of passage of time in a system depends upon the nature of the system itself.

In general relativity, effects due to motion are added to the gravitational effects. Curiously enough, there are some inappropriate logical methods which lead to an accurate result, and explain the slowing-down of clocks in a gravitational field.

A photon has a certain amount of energy, and in view of Einstein's famous formula $E = mc^2$ it is permissible to speak of a photon's 'mass'. According to quantum theory, the change of potential energy of a photon moving out of a gravitational field results in a change of its frequency, and the calculated frequency change is in perfect accord with the predictions of general relativity. This result can be applied to the Schwarzschild singularity, and there is full agreement with the values of the critical radius as listed in table 18. Note, in passing, that some authorities have tried to explain Langevin's paradox by supposing that an observer who experiences acceleration in leaving the Earth will experience the same amount of deceleration on the return journey, so that what he gains on the swings he will lose on the roundabouts!

One of the most important results of general relativity is the equivalence between inertial mass and gravitational mass. Because of this equivalence, it can be said that when a braking action is applied to the projectile in which Langevin's observer is travelling, so that the projectile can return to Earth, then everything takes place as if the projectile had been placed in a potential gravitative field extending from the Earth to the projectile. Under these conditions, the paradox is effectively resolved, and the slowing-down of clocks in the braking field exactly compensates for the time-contraction due to the outward and return motion of the projectile. Unfortunately, this interpretation depends on the idea that events take place as though an attracting field existed everywhere between the Earth and the projectile, and this is not valid. The projectile is really a rocket; it is slowed down, at least theoretically, because it carries its own braking system. From the mechanical viewpoint, the Earth does not influence things which take place at a distance of several light-years, so that Langevin's paradox remains.

Obviously, great care is needed in trying to define the relationships between the local times of two different observers. Logically, the relationships can be stated, but in cosmology, a new fact must be considered: the unity of the universe. When time is defined from one point to another, with reference to the time at a selected point, a unique time can be defined for the universe as a whole. This unique time can be identified with the local time of observers in our Galaxy, and is then called the cosmic time. The evolution of the universe, and all the physical variables which are found in the universe, can then be described by means of this cosmic time.

In part 3 it was noted that time can be defined in various ways; there is mechanical time as well as energy time, electromagnetic radiation time, and radioactivity time. The fundamental postulate is that 'time' as defined by any of these phenomena is always the same, so that the chronology of a phenomenon C is the same whether it is fixed by the time scale of phenomenon A or by that of phenomenon B.

Ernst Mach, one of the founders of modern mechanical ideas, gave a very subtle discussion of the problem of mechanical time. He suggested that the inertia of a body is not independent of other bodies, so that in effect the inertia of a body ought to be defined in terms of the distribution of all the other masses in the universe. Mach's idea, taken together with the identity of inertial mass and gravitational mass, means that the distribution of masses in the universe ought to fix the law of interaction; that is to say, it ought to determine the value of Newton's constant of gravitation.

This was the programme that Einstein proposed when, with general relativity, he put forward a new theory of gravitation. He also noted that his programme had not been fully carried out. The law of attraction between bodies remains tied to a fundamental constant, and does not depend on the distribution of all the other masses in the universe. It is understandable that some physicists have considered going back to Mach's idea and Einstein's programme in an effort to work out a new theory of gravitation, in which both inertia and attraction would depend on the distribution of all the masses in the universe.

In a homogeneous universe, in which the same properties would be valid for every point, the distribution of masses of matter in the universe would depend only on cosmic time, so that both inertia and the mutual attraction of bodies would also depend upon it.

Locally, the significance of this would be that the constant of universal gravitation would not be a genuine constant at all, but would depend upon cosmic time.

The cosmic horizon

Red shift increases with distance. In every model of the universe there must be a distance beyond which the shift becomes infinite, so that wavelengths also become infinite and objects disappear from view. The distance at which this happens is called the cosmic horizon, and marks the limit of the observable universe. The effect applies to a closed spherical universe just as strongly as to a universe which is hyperbolic and open.

The cosmic horizon can be described according to various models, and it will be as well to start with our model of the universe in which there is no cosmical repulsion. For this model, it can be calculated that the line of the cosmic horizon lies at a distance of 9,530,000,000 light-years. The value of the red shift as a function of distance can be calculated, leading to the results given in table 20. The table also shows the apparent wavelengths of the K line of calcium, the violet magnesium line, and the remote ultra-violet hydrogen line when corrections to allow for the red shift have been made. Data of this kind are of more than academic interest. With some of the quasars, the red shifts are great enough to bring the normally remote ultra-violet lines down into the visible range. As an example, consider one particular quasar which has been identified fairly recently. The 2,802 Ångström line of magnesium has been shifted toward the red as far as 7,500 Ångströms. The apparent velocity of recession is 510,000 kilometres per second; this is not, of course, the real velocity, and in a hyperbolic universe the true rate of recession has been worked out at 240,000 kilometres per second. This means that the quasar must be about 6,000 million light-years away.

An even more remarkable case is that of the quasar 3C-9, for which the red shift is approximately 2, corresponding to an

Table 20 Red shift as a function of distance (open universe, no cosmical repulsion).

Distance (thousands of millions of light-years)	**Red shift**	**Apparent wavelength:** K (calcium)	Mg II	Lyman-α
0	0	3,933	2,802	1,215
1·90	0·252	4,900		
3·81	0·668	6,520	4,650	
5·72	1·5		7,000	3,300
7·62	4·1			6,190
9·53	infinite			

Table 21 Fraction of the universe which is observable (spherical universe, H = 100 km/sec/megaparsec).

Density (gm. cm.$^{-3}$)	**Observable fraction**
$2{\cdot}4 \times 10^{-29}$	0·005
$3{\cdot}6 \times 10^{-29}$	0·018
$7{\cdot}2 \times 10^{-29}$	0·051

apparent recessional velocity of 600,000 kilometres per second. On the hyperbolic model, all the calculations fix the cosmic horizon at a distance of about 9,500,000,000 light-years.

Even in a closed universe, with a positive curvature, there is a limit beyond which no object can be seen. The observable universe is a very small part of the total universe, as shown by table 21. In this table, the extent of the observable universe has been worked out according to three different density estimates. It seems that beyond the observable universe there must be still more matter whose amount is infinite in the open model, very large in the case of a closed universe. One of the features of cosmology is its aim of using the properties of visible matter to describe those of matter which is invisible and which will, no doubt, remain forever beyond our reach.

Continuous creation

If we accept the evidence of the observed density of the universe and the value of Hubble's Constant, the universe is about 10,000 million years old. This does not seem very long in comparison with the probable age of our Galaxy and the oldest star-clusters in the Galaxy, but to suppose that the Galaxy is older than the universe is obviously absurd, and somewhere or other there must be an error in interpretation. One solution is to suppose that the gravitational constant varies over a very long period of time, but a completely different explanation was put forward about 1950 by H. Bondi, T. Gold and F. Hoyle, all then working at Cambridge University. This was the so-called 'perfect cosmological principle', involving a steady-state universe.

According to this theory, there could be no difficulty about very great ages. In fact, the time-scale could be pushed back indefinitely, since the Cambridge astronomers supposed that the universe has existed forever, with the same large-scale mean density that it has today; there was no beginning, and there will be no end. The density of the universe considered as a whole never changes, and

Table 22 Number of galaxies inside a sphere of radius 10^8 parsecs, as a function of their age (steady-state theory).

Age in thousands of millions of years	Number
10	$4{\cdot}2 \times 10^4$
20	$2{\cdot}1 \times 10^3$
50	0·25
100	$8{\cdot}5 \times 10^{-7}$

new matter appears continuously so as to compensate for a decrease in density due to the general expansion. The quantity of matter created spontaneously in this way is extremely small, of the order of the mass of one atom of hydrogen per cubic metre per ten thousand million years, but it would suffice. It was stressed that absolutely nothing is known about the form in which matter is created, and there can be no physical theories, even tentative in nature, to account for such a process. The hypothesis of a steady-state universe leads on to the assumption that new galaxies are constantly being formed in a universe in which very old systems already exist.

Let us look more closely at the consequences of the 'creation' of new galaxies in a universe that is expanding. Galaxies 'created' at any particular epoch are now to be found scattered through a volume of space much greater than the volume at the time of their creation. With a creation rate of 5×10^{-48} grams per cubic centimetre per second, it is possible to calculate the number of galaxies of different ages to be found today inside a sphere of radius 100 million parsecs. The results are given in table 22. The number of very old objects to be expected is therefore very low, and such objects would be extremely difficult to detect, for nothing is known about the evolution of a galaxy over so great a period of time.

Gold, for example, proposed an explanation for the quasars, which he thought to be due to frequent encounters between stars in a very old galaxy which had contracted to a small volume, so that the component stars would be relatively close together, and encounters would be common. However, the time needed for a galaxy to evolve to such a state is extremely long, and such an end to the evolutionary process would not be consistent with the idea of a steady-state universe.

A possible test would be to find out the ages of the oldest galaxies, using methods of the same type as for those for fixing the ages of star-clusters, but unfortunately it is not possible to estimate the ages of the galaxies at all accurately, so that the tests available at the moment are inconclusive. Another test would be to study the variation of Hubble's constant over periods of cosmic time. With a steady-state universe, the famous quantity q_0 would be not zero, but equal to -1; but, as has been noted earlier, the observed value of the parameter q_0 does not seem to be compatible with the value of -1 predicted by the steady-state universe theory.

Comment about time

During the course of his investigations on physical subjects, Dirac, in 1939, considered a number of dimensionless magnitudes which were expressed in terms of very large numbers. For example, he considered the relation between the electrostatic force of attraction between the proton and the electron, and the gravitational bond between the two particles. The relationship between these two forces is of the order of 10^{40}. In the same way, Dirac noted that the relationship between the radius of curvature of the universe and Hubble's constant is also in the region of 10^{40}. He concluded that because of some new physical principle, the two values were the same for some definite reason, and not by sheer coincidence, and that because the radius of curvature of the universe varied with time, on account of expansion, the constant of universal gravitation must also vary with time.

This idea was re-examined by Jordan in the years following 1939, using more refined techniques, and more recently Dicke has returned to it. The result has been a new theory in which the constant of gravitation really does change over long periods of time, and this in turn involves a modification of the fundamental principles which give the gravitational equation in general relativity.

The search for proof

In any cosmological theory, observational proof is naturally of the greatest importance. In Jordan's theory, it is assumed that the constant of universal gravitation varies with time, and that this also applies to certain atomic constants. The immediate result is that some other phenomena which depend on the properties of the atom would also have to vary with time, in particular, the wavelength of the 21-centimetre hydrogen line. Observations of the 21-centimetre line are not precise enough to show whether this is the case or not, but it certainly seems most improbable.

Of course, other theories might provide for a variation of the gravitational constant without involving any change in the atomic constants, and this is what is found in Dicke's hypothesis. Dicke set out to discover the effects of a variation in the gravitational constant upon various objects whose evolutionary sequences are known reasonably well; that is to say, stars in the star-clusters; the dynamical evolution of the clusters themselves, and the internal structure of the Sun. Tests based upon the dynamical evolution of clusters are as yet too uncertain to be of much use, but the study of stellar evolution might be more promising. If the evolution of a star took place during a steady variation in the gravitational constant, the ages of clusters would be quite different from the ages usually estimated.

Using Dicke's theory as a basis, Schwarzschild has studied

different models of the internal structure of the Sun to try to find out just what the effects might be, and it is certainly true that classical theory has provided enough information about the age of the Sun and the Solar System to make this investigation worthwhile. Schwarzschild has found that acceptable models of the Sun can be constructed allowing for a slow variation of the gravitational constant, but a variation proportional to the radius of a curvature of the universe would be too rapid for an acceptable model to be worked out.

It is bound to be very difficult to detect a change in the gravitational constant, even if it does occur. The only possibility seems to be to examine the oldest known objects in the universe, for which conventional theories already give plausible estimates; a cluster such as Messier 3, for instance, seems to be as much as 20,000 million years old. The effect of the variation of the gravitational constant would be to speed up the evolution of the stars, so that the new theory would lead to age estimates less than those obtained from theories in which the gravitational constant is taken to be invariable. One interesting result is that the ages of star-clusters would be brought down to below the age of the universe as estimated on a scale based upon the period during which expansion has been going on.

It is dangerous to be misled by seductive results of this kind. Theories of stellar evolution lead to results which are qualitatively of the highest interest, but all these results are subject to revision, and in any case they depend on what we find out in the future about the physical phenomena which control the internal structure of a star.

Several revisions of the estimates of cluster-ages have been made during the last few years, and there is nothing to prove that conventional theories cannot yield results compatible with the hypothesis of expansion.

Nature of the red shift

Little has yet been said so far about the nature of the red shift, which is clearly of vital importance in any discussion. A red shift is, of course, known in the form of a Doppler effect, when a luminous source is moving away from the observer, but general relativity adds two extra reddening mechanisms. One of these is the reddening of a photon which is leaving the gravitational field of a star; the other is an interaction of light with a gravitational field, and is associated with the variation of the radius of curvature of the universe over a period of time.

Over the past forty years, F. Zwicky, who has carried out all his observations at Mount Wilson and Mount Palomar, has questioned the usual interpretation of the red shift, and has always proposed the alternative theory according to which a photon becomes reddened during its movements in space. Physicists have looked critically at this viewpoint, basing their interpretations on ordinary quantum theory. Certainly protons moving in space may well collide with other photons, electrons, neutrinos or other kinds of particles.

In the so-called Compton effect, the frequency of a photon may be altered, and its wavelength lengthened, as a result of an interaction between an X-ray photon and an electron. Therefore it seems reasonable enough to suppose that when a photon suffers a series of collisions with other particles, its wavelength may be increased, which would result in a red shift. Unfortunately there is another point to be borne in mind as well as the conservation of energy: the momentum (product of mass and velocity) must also be conserved. This means that as the photon undergoes its series of collisions, it not only suffers a change in wavelength, but is also affected by diffusion. It is found that the change in wavelength would not only involve a broadening of the spectral lines comparable with the actual change in wavelength, but would also make the images diffuse, and this simply does not happen. The spectral lines of distant galaxies are no broader than those of closer galaxies, and the images are certainly not diffuse. This straightforward fact of observation has always been one of the main reasons for rejecting Zwicky's interpretation of the red shift.

Certainly there is no proof that the suggested process occurs in nature, and this is a definite weak link in the argument; yet it does not seem reasonable to exclude, *a priori*, the theory that the red shift is due merely to an alteration of photons due to new kinds of interactions between photons and other features of the universe. In fact, the idea cannot be rejected out of hand. Incidentally, some years ago it was believed that in a distant galaxy, the 21-centimetre line had been observed to be lengthened in wavelength in the same way as the visible lines; but later work showed that this observation must be erroneous. When it is claimed that red shift is independent of wavelength, all that is really meant is that the shift is independent of the wavelength so far as the visible spectrum is concerned.

The hypothesis of the ageing of photons leads back to a consideration of the steady-state universe, because it involves rejecting the theory that the density of the matter can change over periods of time. This in turn affects all discussions about the evolution of the galaxies, and observations of old galaxies. In such a theory, it must be agreed not only that fresh galaxies are being formed at the present time, but that there is an evolutionary process resulting in the destruction of old galaxies. Therefore, the number of very old galaxies that can be observed ought to be much larger than in the steady-state theory. This is one indication of the difficulties that have to be faced in trying to construct a model for the universe in which the red shift is due simply to the ageing of photons.

Cosmological tests

All sorts of models of the universe have been put forward since the publication of Einstein's theory of general relativity. In some of these, such as Zwicky's, expansion is rejected; Bondi, Gold and Hoyle have favoured the steady-state model; Dirac, Jordan and

Dicke suppose that the gravitational constant varies with time. Few definite conclusions can be reached as yet, but it will be useful to look more closely at the possible tests. As a start, we have various observed facts:

1 The red shift of the spectral lines.
2 The apparent magnitudes of galaxies.
3 The apparent diameters of galaxies or clusters of galaxies.
4 The numbers of objects as a function of their characteristics (or, to be more precise, counts can be made separately for normal galaxies, radio sources and quasars).
5 The study of radio noise in the 3 to 7 centimetre band, which provides information about the temperatures of electrons situated at great distances from us.

Conventionally, a *distance* may be defined as a quantity such that the luminosity falls off according to the inverse square of the distance. To determine this *conventional distance*, the modifications caused by the red shift must be known. This means that the distribution of energy in the spectra of the galaxies must be measured very accurately; the exact correction due to the red shift must be calculated, and the properties of the photographic equipment used must also be taken into account. Certainly the process is not easy, and there are great difficulties in trying to link conventional distance with red shift. So long as the distances remain small, the problem can be solved; but in a curved space in a state of expansion, the red shift is not exactly proportional to the distance, and it may be necessary to see whether terms proportional to the square or the cube of the distance should be introduced. In an expanding universe, the term proportional to the square of the distance is proportional to the acceleration of the expansion, and this may help in deciding whether the rate of expansion is accelerated (positive acceleration) or retarded (negative acceleration). One trouble about trying to apply a correction to the magnitude is that photometry of galaxies is by no means yet perfect. Still, it seems that the expansion is retarded over a time-scale of the order of 2,500 million years. In a static universe this quantity would be infinite; in an Euclidean universe in which the red shift is attributed to the ageing of photons, the expansion would seem to be retarded over a time-scale of 10,000 million years. In short, the measured values for the retardation do not agree either with the idea of a static universe or with the theory of photon-ageing – and other methods of investigation must be sought.

Recent studies of radio noise in the very short wavelength region have led Hoyle and others to reject the steady-state theory. The only logical explanation for the intensity of radio emission at 3 centimetres is that electrons at a very high temperature are seen very far away but with such a red shift that they look very cold. These electrons represent what we presumably see of the very dense universe at the start of the expansion; at that epoch, the matter must have been extremely hot. Calculations have actually shown that models of expanding universes are consistent with the centimetre-wavelength emissions which are observed.

It is probably fair to say that by now the steady-state theory, at least in its original form, has been abandoned by almost all authorities. It was an attractive and plausible idea, but it did not fit the facts, and, like many other attractive theories, it has had to be given up. This does not mean that the 'big bang' idea is necessarily correct, but it does mean that the universe is in some sort of state of evolution.

Quasars

Quasars, as we have seen, could be remaining 'pockets' of the super-dense matter of the universe in its original state; this explanation is perhaps rather like that accepted by Hoyle when he abandoned the steady-state theory.

To justify the steady-state theory, Hoyle had previously been led to a generalisation of the gravitational equations of general relativity; this generalisation could have been interpreted as a continuous-creation term. However, a different interpretation is more likely, according to which the term would be interpreted as a cosmical repulsion term with its value dependent upon the density of matter. Under these conditions, there are possible models in which the high density is accompanied by a repulsion sufficient to allow the model to expand. In these models there are no singularities in the past, that is to say, no stages in which all the matter in the universe is concentrated into a theoretical point, and Hoyle, with his colleague J. V. Narlikar, has described how great masses of matter might first contract and then expand again. This oscillation of very massive bodies, with a stage of minimum radius, could result in the production of quasars.

Matter and anti-matter

When Dirac worked out the theory of electrons, in 1929, he was surprised to find electrons with positive energy as well as those with negative energy. From the physical viewpoint this result was somewhat obscure, and Dicke interpreted these electrons as having a charge opposite to those of normal electrons, so that the energy must be positive. At that time the idea of a positive electron was theoretical only, but it was soon confirmed by actual observation; positive electrons were discovered by Anderson. Nowadays they are generally known as positrons.

When an electron and a positron meet, they annihilate each other with the production of a gamma-ray, and we have here an example of ordinary matter (the electron) meeting what has become called anti-matter (the positron). Apart from the photon, all elementary particles have their 'opposites' as anti-matter; it has already been noted that the anti-proton, for instance, was discovered at Berkeley in 1955. In annihilating each other, the proton and the anti-proton give rise either to pairs of particles and anti-particles, or else to high-energy gamma-radiation.

Nothing in modern theories of elementary particles suggests the stability of anti-particles is any different from that of ordinary particles. The world of anti-particles is just as stable and real as the world of particles. H. Alfvén has suggested introducing a new name, 'koinomatter', for ordinary matter of the kind making up our own part of the universe.

There are two possible hypotheses. According to one, there is a real fundamental dissymmetry between particles and anti-particles, and our universe is made predominantly of particles simply because it is not suited to the formation of anti-particles. Physicists are faced with the problem of trying to work out how stability could be achieved under such circumstances. On the second theory, particles and anti-particles are formed in equal numbers in the universe, and various individual regions happen to be made up either of ordinary matter or anti-matter simply because of chance local conditions. According to the second idea, Klein proposes that bodies in the universe are formed under conditions of equilibrium and high temperature, and that they are produced from a mixture of radiation, particles and anti-particles within their own gravitational field. Only the cooling of massive 'mixtures' of this sort could result in the formation of actual objects.

If these initial conditions are accepted, the next step is to see how a 'galaxy' of anti-particles could exist inside a universe made up of ordinary particles (or koinoparticles, to use Alfvén's term). Alfvén looked critically at this problem, and concluded that between a 'galaxy' of anti-particles and the universe of particles there must be what may be called a frontier zone, where there is an extremely energetic source of radiation which repels the particles and so separates them from the anti-particles. This layer would prevent any catastrophic mixing, and would make encounters between matter and anti-matter a very gradual process. Alfvén has also tried to account for the quasars, which he regards as being due to the emission of radiations and particles of tremendous energy in a stable frontier zone, where matter and anti-matter meet.

The Soviet chemist Semenov has provided a graphic picture of the properties of matter and anti-matter by pointing out that a man made up of matter and a woman composed of anti-matter could meet in space, exchange signals, know each other and even love each other, but they could never touch each other! All this is highly intriguing; but as yet there is no definite evidence either for or against Alfvén's theory.

The question of anti-matter also crops up when we consider the steady-state universe. With continuous creation, matter could well appear spontaneously together with equal amounts of anti-matter. But whether material appeared in the form of neutrons and anti-neutrons or of protons and anti-protons, the meeting of a particle and an anti-particle would still be followed by the emission of highly penetrative gamma-rays, and it is possible to calculate what the flux of gamma-radiation reaching the Earth should be. Here an observational check can be made, because the American satellite Explorer XI detected gamma-radiation in the neighbourhood of the Earth. It seems that the flux is a million times less than the flux that would be required by the theory, and so it seems safe to say that even if continuous creation is going on, it does not involve the production of equal numbers of particles and anti-particles.

The ages of the stars and the universe

It has already been noted that the ages of star-clusters can be estimated, and that studies of radioactive elements lead to at least an approximate determination of the epoch at which the clusters were formed. The oldest objects are perhaps over 10,000 million years in age, and these estimates, rough though they may be, seem to agree quite well with the time-scale of 10,000 million years deduced from Hubble's constant. Of course, if firm evidence were forthcoming that some of the objects in the universe must be older than 10,000 million years, the whole questions of the nature of the expansion, the red shift, and the structure of the universe would have to be critically re-examined.

The diameters of the clusters of galaxies

Zwicky suggested that in order to avoid having to correct the magnitudes of the clusters of galaxies due to the red shift, it would be helpful to count the clusters to draw up a relationship between their numbers and their diameters. Assuming that all clusters are of much the same size, and that they spread throughout the whole of space, the relationship would be quite a simple one, but would be different for an expanding universe than for a static one.

In theory, this is all very well – the test would be better than one involving magnitudes. Unfortunately it introduces the effects of observational selection, because very faint clusters would probably escape detection altogether; the numbers of clusters of small apparent diameter would be too low, and the clusters of large diameter, which would certainly be observed, are not numerous enough to provide the data for a reliable statistical analysis. Zwicky's curve suggests that there is no expansion, but his numbers can be differently interpreted and so made to agree with an expanding universe. More information is needed before the test can be properly carried out.

Counts of galaxies

Moreover, Zwicky's hypothesis is over-simplified, because it is known that the sizes and populations of different clusters are not uniform. Studies of the numbers of galaxies as a function of magnitude should, naturally, take heed of the statistical properties of the distribution of galaxies in a cluster. As an analogy, let us suppose that we are observing a large forest from a distance, and that the forest contains trees of all ages, arranged in groups. The important factors are first, the number of trees in each group; and secondly, the distribution of the trees in different age-groups. Straightforward counting of the number of trees should provide information about the groups, provided that the law of distribution by age-group is known.

Analogies of this sort show that the general problem is by no means simple. And with the galaxies there are many things to be taken into account; just as a small, nearby tree could easily be confused with a large, distant tree, so a near, faint galaxy is only too easy to confuse with a galaxy which is much brighter and much further away. To try to find out the space-distribution of galaxies from their visible aspect is rather like disregarding the colours of leaves which will distinguish a young tree from an old one. Also, just as the slope of the ground could affect the statistical analysis in the tree example, because the number of trees per unit surface must depend upon the slope of the mountain, so the statistics of galaxies are bound to be somewhat affected by expansion. However, the results obtainable are confined to an area so close to the Earth that the effects of expansion would be too slight to be noticed, and here, too, the test is inconclusive.

Counts made separately for normal galaxies, radio sources and quasars seem to be more promising, because the numbers involved are fairly large (around 1,200) and because there are some very brilliant objects which can be seen across tremendous distances. Counts of radio sources, undertaken mainly by Ryle and his colleagues at Cambridge, do not support the steady-state theory, though it is true that the effects of observational selection are still not known as well as might be hoped.

The Singular State, or the Big Bang

Homogeneous, isotropic models of the universe all suppose that the expansion must have started from a 'singular state' of infinite density, pressure and temperature, so that there has been much speculation about the physical processes operating at the moment of the so-called 'big bang'. Actually, the original singular state could be dispensed with if the models are homogeneous but not isotropic, because such models would show 'slipping', slipping with rotation, or rotation alone. Therefore, the singular state has changed its nature, and the density, pressure and temperature need not be taken as being infinite. Even a slight departure from isotropy is enough to cause a complete change in the character of the singular state through which the universe has passed, and the departure is even greater when rotation is taken into account. It then seems that the initial 'big bang' does not account for the expansion now in progress. There is no need for the modifications of the equations of general relativity, as suggested by Hoyle and Narlikar, to avoid a physical singular state at the cost of the expansion.

This may be a suitable moment to say something further about general relativity. In the theory of gravitation, it is not really natural to introduce *a priori* the distribution of matter in space. The more natural procedure, as followed by Einstein, Infeld and Hoffman, is to suppose that matter acts as though it were a distribution of mass-singularities in space, so that there is no need to introduce, artificially, a continuous distribution of matter in space; it is enough to study the relationships between attracting 'singular points'. When such a relativistic system is studied to successively higher approximations, different properties come to light in succession. When the approximations are rough, the material singular points are displaced with uniform motion, or, in mechanical terminology, with Galilean motion. Newton's law of universal attraction, and the fundamental dynamical principle according to which force is equal to the mass multiplied by acceleration, are discovered at the next more refined approximation; and at the next approximation still, the first relativistic corrections are found, as in the two-body problem corresponding to the advance of the perihelion of the planet Mercury. The importance of this analysis is that it stresses the link between the movements of material bodies and the laws of dynamics. Everything hinges upon the bond between the singular field of each particle and the fields of all the other particles.

Applying these ideas to the universe is one of the most important tasks of general relativity, and brief mention must be made of the theory put forward by Pachner, which is very difficult from a technical point of view but which ought certainly to be verified and developed. What has been said here should serve to emphasise that even in its conventional form, general relativity has by no means realised all its possibilities as yet.

The universe and cosmogony

Undoubtedly the present time is a critical one in cosmology. With the advances in studies of quasi-stellar radio sources, extraordinarily powerful means of sounding the depths of the universe have been discovered, and the problem of origins of these sources, a matter of vital importance in *cosmogony*, is closely allied with the *cosmological* problem of the structure of the universe. It is not impossible that the whole question is linked with the whole manifestation of the material in the universe.

In any case, studies of quasars should soon be able to provide answers to the outstanding riddles concerning the expansion of the universe. What has to be done is to probe as far into space as is possible, and for this the incredibly brilliant quasars hold out much the best hopes.

Conclusions

The universe is made up of stars which are grouped together in galaxies of innumerable shapes and forms. The average density of the material available to observation is very low, of the order of three atoms of hydrogen per ten cubic metres, but this material makes up objects of enormous complexity. In every direction, as far as we can see, there are galaxies of stars. The stars are formed constantly, and go through processes of evolution; the galaxies, too, are born, develop and die. But while a great deal has been learned about the evolution of the stars, our knowledge of the evolution of the galaxies is still depressingly meagre.

The great galaxies, as luminous as tens of thousands of millions of suns, can be seen across distances of thousands of millions of light-years. A hundred times brighter still are the quasars, which bring us to the limit of that part of the universe which is within our observable range. Nuclear sources can account for the energy of normal galaxies, but this is not so for the radio galaxies, while the quasars must be drawing upon some mysterious energy-supply which can give them their tremendous brilliance and also produce an abundant flux of particles which seems to resemble the particles of cosmic radiation.

Whether the universe is spherical or hyperbolic, observations of the galaxies force us to abandon Euclidean geometry in our efforts to understand the strange properties of the curvature of space. What is termed Riemannian geometry, not discussed here because it is beyond the scope of the present book, has become a physical reality instead of nothing more than an abstract mathematical hypothesis. But though we may assume that the universe is curved, the basic concepts still elude us. The newly-found very blue galaxies, the quasi-stellar sources, are so far off that the measures made of them are highly relevant in this connection; the most remote have red shifts so great that the Lyman-alpha line is observed in the blue part of the spectrum.

Quasi-stellar sources are being discovered at a surprising rate, and dozens of them are now known. Their mean absolute magnitude seems to be $-24{\cdot}7$, corresponding to the retarded expansion of a closed spherical universe in which the parameter q_0, a measure of the retardation, works out at $+1$. It is quite likely that over 100,000 quasars above an apparent magnitude 19 exist, and the analysis of the very blue objects referred to earlier in part 5 is even more interesting, as well as perhaps more deceptive, than it appeared to be a year or two ago. Yet counts of them cannot be a reliable guide to the properties of cosmological models, and until spectroscopic analysis of their light become possible it will be very difficult to come to any conclusions about their true nature. Such an analysis will be a difficult matter, and will take a long time. On the other hand, it has been shown that the numbers of very blue objects, that is to say, quasi-stellar sources, are very large, and must be measured in hundreds of thousands.

Another observation made comparatively recently may also turn out to be fundamental – the detection of background radio emission from the sky. The research has been carried out at wavelengths of 2·4 centimetres, 3·3 to 7 centimetres, and 21 centimetres by the Bell Telephone Company and the Princeton Laboratory; the intensity seems to correspond to the thermal agitation of electrons in a temperature of 3 degrees absolute. This emission does not correspond to any known radio sources, and the measures seem to be accurate to within 20 per cent.

According to one interpretation, the electrons of intergalactic space are too few in number to contribute appreciably to the general background radio emission, even if this were probable on other grounds, which it is not. Therefore, the cause lies in some very distant source, possibly connected with the primitive state of the universe immediately following the start of the expansion; in fact, the thermal emission represents the remains of the big bang. Efforts have been made to develop as coherent a theory as possible, accounting for the background radio emission in the centimetre wave-band as well as other known facts. The original expansion of a very hot, dense universe presumably took place very rapidly, and after some thousands of years led to the residual gas made up of cold electrons. This seems to account for the background emission, but it may be possible to consider events of an even earlier period, and this has led to some sort of a revival of an old theory put forward by Alpher, Bethe and Gamow to explain the origin of the various chemical elements.

When the age of the universe was 1/100 of a second, then, according to Alpher and his colleagues, the temperature was of the order of 100,000 million degrees centigrade, with a density of 1,000 grams per cubic centimetre. As rapid cooling took place, different kinds of stable nuclei were formed. Calculations can be made for the abundance of helium and deuterium at the virtual start of the expansion, because we know the present density of the material in the universe, and we also know the present abundance of helium and deuterium, as well as the temperatures of the electrons near the boundary of the observable universe. So far as helium is concerned, initial densities of 2×10^{-29}, 7×10^{-31} and 2×10^{-32} grams per cubic centimetre are all compatible with the available data. The results for deuterium are much less satisfactory, because so much depends upon the mean density of the material in the universe; the hypothesis of 2×10^{-32} grams per cubic centimetre leads to a present abundance for deuterium of about 1 per cent, which is about fifty times too high. A value between 2×10^{-29} and 7×10^{-31} gives much better agreement, but no final conclusions can be drawn, because it is known that the material now spread through our Galaxy has passed through the stellar state several times, during which periods deuterium must be destroyed in large quantities.

The Solar System is probably less than 5,000 million years old, and so is much younger than the Galaxy, whose age is certainly 10,000 million years and perhaps more. During the 5,000 million years which elapsed between the beginning of the Galaxy and the origin of the Sun, the material may well have been contained partly in stars and partly as interstellar matter. Mixing of this sort is bound to result in considerable losses of deuterium. It is notable that the abundance of deuterium found on Earth and in meteorites is about 1/10,000 of the abundance of hydrogen, and this may be the result of deuterium having been destroyed in the course of its period within the interiors of stars. In any case, the results seem to favour a model of a closed spherical universe, with the background radio noise originating from regions so far away that the symbolic velocities of recession amount to several thousands of millions of kilometres per second.

The analysis of the colour/absolute magnitude diagram for globular clusters, taken together with the inner structures of the clusters themselves, ought to shed some light on the probable abundance of primitive helium at the time when the galaxies were formed. There is a good deal of uncertainty, because it looks as though different globulars have different amounts of helium at the present time, and some of them are helium-deficient. This sort of deficiency would favour open, hyperbolic models of the universe with a low density for the material, but the destruction of primitive deuterium must also be borne in mind.

To sum up: Either primitive helium was relatively scarce (with abundance about 10 per cent) and primitive deuterium has been burned up during the evolution of the stars in the galaxies, so that the evidence favours an open, hyperbolic model for the universe; or else primitive helium was abundant (around 30 per cent) and the present abundance of deuterium is the same as it has always been, in which case the universe is closed and spherical.

Counting methods seemed formerly to hold out hopes with regard to the determination of cosmological models, but have proved to be something of a disappointment, because we have no real idea of how the galaxies have developed over long periods of time. Analyses of radio sources also seemed promising, because the sources are powerful enough to be detected over great distances, and yet quasar counts do not seem to fit in with any cosmological models so far proposed. It looks as though either the numbers of quasars evolve in time, or else the quasars themselves evolve with time. If so, quasars may provide vital clues to the overall processes of evolution.

Systematic discussions of all homogeneous, uniform models of the universe have been completed by now, so that apart from the expansion parameter and the famous cosmological constant it ought to be possible to work out a parameter expressing rotation and also a parameter expressing the rate of deformation. Investigations of these various models, all of which are based on Einstein's general relativity, confirm that the introduction of rotation or anisotropy raises difficulties when we come to consider expansion from a singular state. Some authorities have questioned the reasons for such investigations; why, in fact, reject the idea of a singular state, when the trouble may lie in the incompleteness of Einstein's equations? After all, the singular state seems to be necessary in explaining the background radio radiation from the sky, and it is worth noting that thirty years ago, Tolmann was speculating as to whether some relic of the universe could be found which would take us back before the singular state, to a time when contraction was still going on.

Moreover, it is also possible that successive oscillations of the universe cause irreversible effects which accumulate – in which case the entropy (i.e. the degree of molecular disorder in a system) of a closed spherical universe should increase steadily. Lively discussions about this point are still going on, and are likely to continue for a long time yet.

Hoyle has now given up the idea of a steady-state universe. If the theory had been correct, matter could never have passed through the physical state which seems necessary to explain the background radio emission. Hoyle has also underlined how his researches into general relativity had led him to attach importance to the terms which prevent the occurrence of a singular state in the universe taken as a whole.

We have seen above that quasars could be interpreted as pieces of matter erupting in our universe. Such views were originally due to Jordan and were taken up again more recently by Novikov and by Neeman. They should certainly be borne in mind, because it is possible with these theories to envisage the appearance of vast masses of matter undergoing differential expansion. Hoyle has attempted to explain the quasars in this way, mainly because of the extraordinary concentration of matter toward the centres of the systems. Authorities who follow more conventional ideas about the quasars are hardly likely to accept new models of this sort. On the other hand, it is true that collisions between stars, with all the consequences arising, could lead to phenomena closely resembling the quasars, provided that the collisions are frequent enough; that is to say, that the stars are sufficiently closely packed. It is not only direct collision that must be considered; we must also remember that the stars will also cause intense friction against nebular matter, and coalescence might result in supernova explosions. The energy of the quasars might be explained in such ways, but it is not easy to see how any systems could have become dense enough for such remarkable interactions between stars to occur.

6 Summary

It is now time to take stock of what has been said in this book. The universe observable from Earth is made up of stars and star-systems, with most systems including tremendous quantities of dust and gas. Our Galaxy is shaped like an immense disc, with a diameter of 100,000 light-years; the Sun lies some 30,000 light-years from the galactic nucleus, and completes one orbit in about 250,000,000 years. There are about 100,000 million stars in the Galaxy; in the neighbourhood of the Sun there is an average of one star in a cube of side ten light-years, and when dust and gas are taken into consideration the average mass works out at two or three atoms of hydrogen per cubic centimetre. It is not possible to estimate the significance of the masses of obscure material, of slight mass, moving inside the Galaxy, but undoubtedly this material must exist.

The chemical composition of the universe taken as a whole is uniform, but tremendous differences in detail are found in different regions, as has been found from studies of the composition of the Earth, the elements found in meteorites, and the relative abundances of the elements as estimated from the spectra of the stars.

In our Galaxy there are many star-clusters, and for many of these reliable age-estimates can be made. Very young clusters appear to be several millions of years old, while others have ages of something of the order of 15,000 million years. Stars evolve over periods of time, and their compositions are altered by nuclear transformations; radioactive elements can be regenerated by the explosions of supernovæ, and it may be that this process began with the formation of our Galaxy 18,000 million years ago.

Beyond our system there are other galaxies, of which one of the nearest is the Andromeda Spiral, at a distance of 2,200,000 light-years. They are of varied forms; elliptical, spiral and diffuse. There are radio galaxies, and also the quasars, often called quasi-stellar radio sources because they show up as starlike points when photographed.

The galaxies evolve. Violent explosions give rise to radio emission, and some galaxies show traces of several successive explosions. Systems of galaxies also evolve; there are multiple systems which seem to be only a few thousands of millions of years old. Quasars, a hundred times more luminous than the brightest elliptical galaxies, are energy sources of unparalleled violence, capable of releasing energy flux of the order of a million million times that of the Sun. Also, the galaxies are spread throughout space, and the brightest objects of all – the quasars – can be measured out to well over 6,000 million light-years. Euclidean geometry is no longer valid for such tremendous distances. The paths of light-rays are modified by the presence of matter, and this leads on to discussions of curved space, where properties analogous to those of two dimensions have to be adapted to a three-dimensional system.

Consistent efforts are being made to decide whether the universe is open and hyperbolic, or closed and spherical. The Hubble-Humason relationship, and the background radio noise from the sky, seem to favour the spherical, closed model, but the matter is by no means settled. The density of material in the universe will be different for different models, but a mean density of 2×10^{-29} does not seem to be incompatible with the best available spectroscopic results. Of course, there are other reasons for uncertainty as well. The expansion constant indicates an age for the universe of about 10,000 million years, so that presumably some extraordinary event (the so-called 'big bang') took place at that time. And yet this seems to contradict the estimates for the ages of the oldest star-clusters in our Galaxy, so that much work remains to be done. A deeper study of the internal structures of the stars will give better values for the ages of the clusters, and these in turn will lead to better estimates for the age of the universe itself.

As yet, nobody knows how these two theories can be reconciled.

Bibliography

If a book has been published both in Britain and the United States both publishers are listed, the British one being named first. Dates are of first publication.

Much of the material in this book is contained in recent issues of technical periodicals, such as the *Astrophysical Journal*, the *Monthly Notices of the Royal Astronomical Society*, the *Astronomical Journal of the USSR* and *Annales D'Astrophysique*. Handbooks containing knowledge at a fairly advanced technical standard include *Stellar Structure*, L. H. Aller and D. B. McLoughlin (eds), University of Chicago Press, London and Chicago, 1962 (volume 8 in a compendium of astronomy and astrophysics entitled *Stars and Stellar Systems*, under the general editorship of Gerard P. Kuiper and Barbara M. Middlehurst).

The following semi-popular books will be found useful:

Abell, G. (1964). *Exploration of the Universe*, Holt Rinehart and Winston, London and New York

Abetti, G. and Hack, M. (1964). *Nebulae and Galaxies*, Faber/Crowell

Alfvén, H. (1967). *Worlds-Antiworlds*, Freeman, London

Baade, W. (1963). *The Evolution of Stars and Galaxies*, Oxford U.P./Harvard U.P.

Bonnor, W. B. (1965). *The Mystery of the Expanding Universe*, Eyre and Spottiswoode/Macmillan

Firsoff, V. A. (1967). *Facing the Universe*, Sidgwick and Jackson, London

Hoyle, F. (1967). *Galaxies, Nuclei and Quasars*, Heinemann/Harper

Jennison, R. C. (1966). *Introduction to Radio Astronomy*, Newnes, London

Kahn, F. D. and Palmer, H. P. (1967). *Quasars: their importance in astronomy and physics*, Manchester U.P./Harvard U.P.

McVittie, G. C. (1961). *Fact and Theory in Cosmology*, Eyre and Spottiswoode/Macmillan

Moore, P. (1967). *Astronomy*, Oldbourne/Grosset and Dunlap

Moore, P. (1967). *The New Look of the Universe*, Zenith/Norton

North, J. D. (1965). *The Measure of the Universe*, Oxford U.P., London and New York

Schatzman, E. (1966). *The Origin and Evolution of the Universe*, Hutchinson/Basic Books

Shapley, B. (1961). *Galaxies*, Oxford U.P./Harvard U.P.

Acknowledgments

Acknowledgment is due to the following for the photographs (the numbers refer to the page on which the illustration appears).

Frontispiece © 1965 California Institute of Technology and Carnegie Institute of Washington; 10 R.A.S., W. Lockyer; 14, 17, 19, 34–5, 36, 38–9, 43, 47, 59, 69, 120–1, 123, 125 Patrick Moore; 22, 26–7, 125 Mount Wilson and Palomar Observatories; 49, 60, 130 R.A.S.; 113 R.A.S., R.O. Cape; 121 Lick Observatory.

Index

World University Library

Books published or in preparation

Economics and Social Studies

The World Cities
Peter Hall, *London*

The Economics of Underdeveloped Countries
Jagdish Bhagwati, *MIT*

Development Planning
Jan Tinbergen, *Rotterdam*

Human Communication
J. L. Aranguren, *Madrid*

Education in the Modern World
John Vaizey, *London*

Money
Roger Opie, *Oxford*

Soviet Economics
Michael Kaser, *Oxford*

Decisive Forces in World Economics
J. L. Sampedro, *Madrid*

Key Issues in Criminology
Roger Hood, *Durham*

Population and History
E. A. Wrigley, *Cambridge*

History

The Emergence of Greek Democracy
W. G. Forrest, *Oxford*

Muhammad and the Conquests of Islam
Francesco Gabrieli, *Rome*

The Civilisation of Charlemagne
Jacques Boussard, *Poitiers*

The Crusades
Geo Widengren, *Uppsala*

The Ottoman Empire
Halil Inalcik, *Ankara*

Humanism in the Renaissance
S. Dresden, *Leyden*

The Rise of Toleration
Henry Kamen, *Warwick*

The Scientific Revolution 1500-1700
Hugh Kearney, *Sussex*

The Left in Europe
David Caute, *London*

The Rise of the Working Class
Jürgen Kuczynski, *Berlin*

Chinese Communism
Robert North, *Stanford*

The Italian City Republics
Daniel Waley, *London*

The Culture of Japan
Mifune Okumura, *Kyoto*

The History of Persia
Jean Aubin, *Paris*

A Short History of China
G. F. Hudson, *Oxford*

The Arts

The Language of Modern Art
Ulf Linde, *Stockholm*

Twentieth Century Music
H. H. Stuckenschmidt, *Berlin*

Art Nouveau
S. Tschudi Madsen, *Oslo*

Palaeolithic Cave Art
P. J. Ucko and A. Rosenfeld, *London*

Primitive Art
Eike Haberland, *Mainz*

Expressionism
John Willett, *London*

Language and Literature

French Literature
Raymond Picard, *Paris*

Russian Writers and Society 1825–1904
Ronald Hingley, *Oxford*

Satire
Matthew Hodgart, *Sussex*

The Romantic Century
Robert Baldick, *Oxford*

Philosophy and Religion

Christian Monasticism
David Knowles, *London*

Witchcraft
Lucy Mair, *London*

Sects
Bryan Wilson, *Oxford*

Earth Sciences and Astronomy

The Structure of the Universe
E. L. Schatzman, *Paris*

Climate and Weather
H. Flohn, *Bonn*

Anatomy of the Earth
André Cailleux, *Paris*

Zoology and Botany

Mimicry in plants and animals
Wolfgang Wickler, *Seewiesen*

Lower Animals
Martin Wells, *Cambridge*

The World of an Insect
Rémy Chauvin, *Strasbourg*

Primates
François Bourlière, *Paris*

The Age of the Dinosaurs
Björn Kurtén, *Helsinki*

Psychology and Human Biology

Eye and Brain
R. L. Gregory, *Edinburgh*

The Ear and the Brain
E. C. Carterette, *UCLA*

The Biology of Work
O. G. Edholm, *London*

The Psychology of Attention
Anne Treisman, *Oxford*

The Psychology of Fear and Stress
J. A. Gray, *Oxford*

The Tasks of Childhood
Philippe Muller, *Neuchâtel*

The Doctor and the Patient
P. Lain Entralgo, *Madrid*

Chinese Medicine
P. Huard and M. Wong, *Paris*

Physical Science and Mathematics

The Quest for Absolute Zero
K. Mendelssohn, *Oxford*

What is Light?
A. C. S. van Heel and
C. H. F. Velzel, *Eindhoven*

Mathematics Observed
Hans Freudenthal, *Utrecht*

Waves and Corpuscles
J. L. Andrade e Silva and G. Lochak, *Paris* Introduction by Louis de Broglie

Applied Science

Words and Waves
A. H. W. Beck, *Cambridge*

The Science of Decision-making
A. Kaufmann, *Paris*

Bionics
Lucien Gérardin, *Paris*

Data Study
J. L. Jolley, *London*